Islands of Connaught

Paintings and Stories from Sailing the Islands

Wallace Clark and Ros Harvey

First published by Cottage Publications,
an imprint of Laurel Cottage Ltd.
Ballyhay, Donaghadee, N. Ireland 2005.
Copyrights Reserved.
© Illustrations by Ros Harvey 2005.
© Text by Wallace Clark 2005.
All rights reserved.
No part of this book may be reproduced or stored on any media
without the express written permission of the publishers.
Design & origination in N. Ireland.
Printed & bound in China.
ISBN 1 900935 47 3

Ros Harvey

Ros Harvey grew up in Malin on Inishowen. She spent her childhood 'messing about in boats' along the spectacular coastline of Donegal.

Ros had been a leading potter in Dublin for twelve years until a back operation stopped her working in that medium. She then moved into painting using soft pastels. She finds the tactile qualities of pastels a natural transition from clay.

The advantage of these pure colours enables her to recreate the fast changing light and atmosphere of this western coast.

She has exhibited her paintings and prints in England, Germany, Holland, Italy and the U.S., as well as at the annual Academy exhibitions in Belfast and Dublin and at the Royal Society of Marine Artists in London.

Islands of Connaught is a natural follow on to *Donegal Islands* and *Inishowen*, earlier best sellers in this series which Ros also illustrated.

Wallace Clark

Wallace Clark is a linen manufacturer by trade and a sailor by inclination. He has been making friends in the Donegal islands by visits in boats varying from currachs to galleys, dinghies and cutters since 1948.

His sailing interests cover the spectrum from leisure sailing to participating in several research projects to explore the seafaring technology of bygone times.

These included a currach voyage from Derry to Iona in 1963, participation on Tim Severin's Brendan Voyage across the Atlantic in 1977 and the Lord of the Isles Voyage when Wallace oversaw the construction of the first Highland Galley in 300 years and her voyage from Westport to Stornoway.

Drawing on this rich experience he has written a number of books about sailing and islands including: *Rathlin – Its Island Story; The Lord of the Isles Voyage; Sailing Round Russia* and, perhaps his best known book, *Sailing Round Ireland* which is regarded as a classic of its kind.

Contents

1. Metal Man
2. Coney Island
3. Doonbristy
4. Puffin Island
5. Pig Island
6. The Buddagh
7. The Stags
8. Kid Island
9. Eagle Island
10. Inishglora
11. Inishkea North
12. Inishkea South
13. Achill Island
14. Inish Biggle
15. Achillbeg
16. Clew Bay

17. Clare island
18. Inishturk
19. Caher
20. Davillaun
21. Inishbofin
22. Inishark
23. High Island
24. Friar Island
25. Cruagh Island
26. Omey Island
27. Inishlacken
28. St. Macdara's Island
29. Masson Island
30. Inishmore
31. Inishmaan
32. Inisoirr

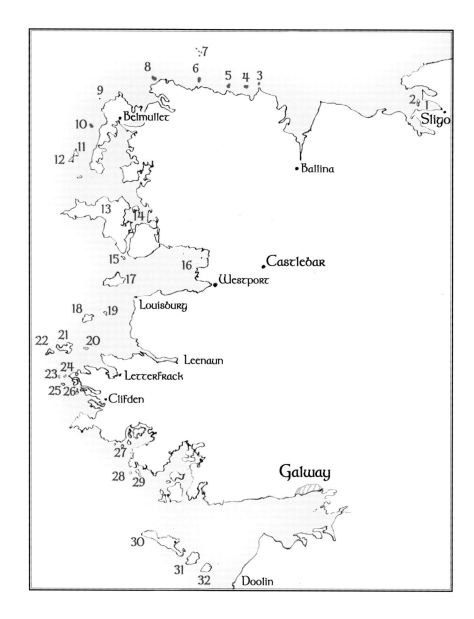

Introduction

'I am haunted by numerous islands
And many a Danaan shore
Where time shall surely forget us
And sorrow come near us no more.'

The White Birds by W.B. Yeats

Ireland is an island, the second largest in Europe. So all of us who are lucky enough to live here are islanders and can share in part the problems and pleasures of those who live on our remoter offshore islets.

Ireland's islands lie round her shoreline like water drops fallen from the head of a swimmer. Some are single, others in little groups thrown off by the swimmer's nose and chin as they turns their head sideways to breathe. Like water drops they glisten in many colours.

The islands each have their own special character. Many are far enough from the coast for there to be a sense of achievement in getting there but all close enough for a weekend visit.

Why do they attract us?

Editors like islands because they make catchy headlines.

Biologists like insular work because they can do a finite survey of plant and animal life. Teams have done this twice at long intervals on Clare Island and discovered lots of species unknown to science. Perhaps it is just as well these are mostly insect-sized. We don't need any mini-dinosaurs just yet.

Sailors like islands for their shelter, ready friendships, fresh water and wildlife.

Tourists like islands to get away from appointments and brown window envelopes. And because they are different and have sparkling waters and private bays that need no blue flag to tell you they are unpolluted.

Island people tend to be happier with their lot and more philosophical than the rest of us. Misfortunes which would make a mainlander bang his head against the wall are shrugged off with, 'Augh! These things happen'. They live in a real world of cause and effect.

My cousin Ros, as an artist, likes islands because of the special overall radiant light. She likes the rugged wild approaches and the anticipation of what lies ahead when the boat slips round a point to a sheltered harbour or cove.

She likes the atmosphere of the past inhabitants and the warm welcomes of the present. Ros thanks the District Nurses on the larger islands, ever-ready to bandage, soothe and administer orthodox and alternative medicines, at any hour, and bid you farewell with crabs and carrageen moss!

I like islands for almost everything but best for re-meeting old friends who live there, including the seabirds and because there is a chance of adventure along the way; more so if you are able to go in your own small boat.

This book which Ros and I have had such fun putting together is about the isles in the middle of the west coast. Connaught has a hundred and fifty miles of coastline pitted and riven where Atlantic storms have played 'ballyhooly' with rocks and bays.

The Scots define an island as any detached piece of real estate with enough grass to support a family or one sheep. Anything smaller is a rock. By that standard there are over a hundred islands between Sligo and Galway. From this we have selected a couple of dozen.

Please don't be upset if your favourite *isola bella* is omitted. There isn't room in a book of this sort to mention them all. Anyway if you want to keep your isle private the less said about it the better.

Harbours giving full shelter on islands are rare. Connaught includes two of the best protected in Ireland, Inishmore in the Arans and Inishbofin. Perhaps it is these and the shorter distances from the Continent that first gave Connaught closer contacts with French, Spanish and Dutch traders and raiders than islands up north.

From outsiders came dangers as well as new ideas. Island forts were neeeded to protect families from pillage and enslavement. More varied architecture came in and monks - lots of them - with strong faith. Connaught sailors from as far back as the 6th century knew more of the Atlantic than other Europeans. Columbus, in 1492, took William Irez of Galway with him as a pilot.

Islands produce strong characters. From Clare Island came Granuaile, anglice Grace O'Malley, the most famous character in Irish maritime history.

It is widely, and scurrilously, believed that sailors have a girl on every island. It took my Russian sailor friend Nikolai Litau to refute such slander.

"Nonsense", he said. "Totally untrue."

"Why's that?"

"We haven't been to every island!"

Confucius knew that a thousand years ago when he remarked, 'The humane man takes his pleasure from the mountains. The wise man takes his delight from islands.'

Let's go and take our delight among plenty of isles with lots of uninhibited inhabitants.

Thanks

Many friends from Connaught have helped with information, advice and corrections.

John and Sheila Mulloy for source material and painstaking proof reading, Margaret Day for breezy updates on Inishbofin, John Coyle, Archdeacon Thomas Stoney, Dr Des Moran, Feichin Mulkerrin, Chris O'Grady and Vincent Sweeney.

The boats which served us so well varied from *Zamorin*, a fifty year old gaff cutter to, *Cura*, a beautiful Mc Gruer five ton sloop, then for forty years, *Wild Goose*, a 35 foot Maurice Griffiths yawl and currently, *Agivey*, a 32 foot Colvic ketch, named after a river near home.

Without loyal crew members our boats would never have reached the isles. Early on there was my wife June, brother Henry and sister Jill, son Milo, Jan Eccles, Alan, Mike, Chris, Willie and Harry.

More recently Ricky, Graham, Lewis and Melanie, Terry and Paddy have helped navigate *Agivey*.

What good crack you all were!

Sailing to the Arans

Preamble

'Turn ye to the Stronghold, ye Prisoners of Hope'.

Zach. 9 – 12

The above quotation sums up why people fled more than 1500 years ago to the Aran Isles in Galway Bay. There they built mighty strongholds to deter attack. These were so strong as to keep the community free for centuries from the worst threat of piracy.

That threat was not imaginary and continued up until the 17th century when Algerine pirates took 20 men plus 89 women and children into slavery from Baltimore on 20th June 1625. They would have thought little of coming on up north for pickings looked easy.

The Arans are fortified more skilfully than any European island or peninsula of their period except perhaps Gibraltar.

Their Christianity was on a similar scale. 'No district in Europe, whether insular or inland, contains in the same space so many monuments of the early Christian Church', a clerical researcher stated in 1862.

The Arans Islands are a wedge in three parts, fifteen sea miles long and scarcely two wide, with cliffs on the south side rising 300 feet. They are the largest and most populous of any Irish group, also the barest – only six per cent of their eleven thousand acres is arable. The fact of being unique in half a dozen different ways is what attracts visitors in ever growing numbers.

Inishmore

'At first it seemed a little speck
And then it seemed a mist
A speck, a mist, a shape, I wist
And then it near'd and near'd'

Coleridge

Our first Aran visit was in *Zamorin* in June 1952. As we drifted gently east from Slyne Head a dozen brown sails appeared on our bow. They were hookers bound south for Aran from Connemara. Soon we could make out their black tarred hulls and turf cargoes piled high above the gunwales.

Currachs were around us fishing on a sea that swelled blue under circling gulls as the hundred foot Eeragh Lighthouse came abeam on Brannock at the west end. It was off here, in a monstrous sea, that the brilliant boat-handling of Paddy

Stone Wall, Inishmore

Mullan was demonstrated for the 1935 film *Man of Aran*. Its success first put the islands on the world stage.

White cottages appeared sprinkled over the grey landscape until six miles further on we turned south leaving Straw Island's neat white lighthouse and beach to port.

Passing the moored lifeboat we tied alongside the massive stones of Kilronan pier. Pat Flaherty, an acquaintance from Glashnacally on the mainland, greeted us, a big man with broad shoulders, a stiff chin and light blue eyes. He wore thick wool trousers, a dark blue jersey under a buttonless waistcoat and cloth cap at a jaunty angle. Older men on the quay affected the traditional Aran pampooties. These are made from half-tanned hide. In old days sealskin was counted best. With hair on the outside it gave a good grip on the rocks when launching a currach.

Deals were being made and Pat's cargo thrown up, turf by turf, onto the quay and into donkey panniers. A return cargo of kelp and Laminaria seaweed for making alginate awaited collection at the head of the pier.

Pat proudly showed us his hooker of a design so old that its origin is lost. She was 35 feet long of massive construction, with a half deck upfront, a bailing-well aft and a wooden firkin on a long pole took the place of a pump. Potatoes simmered on a slate hearth over glowing turfs.

We set off west at a canter by sidecar for Dun Aengus, the grandest fort of all. In these conveyances passengers face outwards. A fall from that position, as the car bucked and swayed like a ship at sea, on the narrow roads with a lime-stone wall on each side would have meant nothing less than a broken nose. We held on tight.

On both sides of the lane lay a chequer board of tiny stone-bound fields.

Pateen, our jarvey, pointed out numerous ruined sites – castles, cottages, churches and oratories. A few stunted trees grew near the shoreline to our right.

To our left on rising ground were acres of flat stone, split as if by a giant with a chisel and smoothed by his ten-foot plane. Stones on Aran make up three quarters of every view.

Dun Aengus sat like a shaved coxcomb on the skyline ahead.

We walked up two hundred yards to pass through an outer wall enclosing ten acres of sloping rock and grass. At its far end lay the intermediate fortification – sharp stones, close set in random array at acute angles. These were enough to slow up infantry of any age and give spear men on the walls time to pick their targets.

Passing into the low narrow entrance we gasped at the sight of the inner courtyard, flat, semi-circular and big enough to hold three tennis courts. More astonishing was its far side – fenceless on the edge of overhanging 200 foot sea cliffs. Pateen claimed the builders as his forbears – Firbolgs, meaning bagmen, who created soil on the Arans by bringing out sacks of earth from Connemara.

Armed only with bronze they were in terror of extinction by recent arrival in Ireland of a race known as Danaans equipped with swords of iron. The Firbolgs construction efforts paid off. There is no record of Dun Aengus in its

Dun Aengus, Inishmore

2,500 year existence being taken by assault. When danger threatened, whole villages moved into the forts and performed every function of their existence between the outer and inner walls.

There were many invaders – Irish freebooters, Dutch fishermen, English and French privateers plundered in turn. They failed to leave a forwarding address so records of their depredations are nonextant.

An elderly man of Aran told me in 1952 that Dun Aengus was taken once by stealth when Vikings climbed the cliffs and overcame the lookouts but no written record survives.

The charm of Dun Aengus lies in three things – its size, its cliff edge site and its powerful landward defences. "Every stone of them", as Pateen pointed out, "laid by the hand of man". There was then time to visit a cluster of roofless buildings a mile west of Dun Aengus called the Seven Churches. Here thousands of students studied the Gospels and learned the art of writing and illumination.

When we got back to Kilronan, Joe Doyle, the Life Boat Mechanic, had fixed our yacht engine which had been unwell and got it going with an expert flick of the starting handle. He waved aside any question of payment, by saying, "The wife has tea for yins at home". There by the quay, griddled plaice, home-made scones with salted butter and strong tea made a meal for heroes.

That is the spontaneous hospitality which makes visiting Irish islands so delightful.

Then we went shopping: potatoes from the butcher, he had a plot; milk from the hardware shop, the owner had a cow; bread from the Post Office, it came with the mail; veggies from the blacksmith who tilled a pretty garden. It was simple but the pleasant Irish tradition that no business can be transacted without some conversation meant it took a couple of hours.

We spent the evening in Killeany a mile south of Kilronan. It overlooks the old part of the harbour, dry at low tide. The ruins of a Cromwellian fort, garrisoned to protect the fishing grounds from the Dutch and French, felt warm in the evening sun.

In a bar seemingly filled by big raw-boned men with pale blue eyes and broad shoulders, friendly islanders all. We were told that Killeany is named after St. Enda whose grave is nearby.

The worst Viking raid was in 1091 – after that the O'Briens from County Clare owned the islands for 400 years; wild men, always fighting, but not seamen.

Then came the O'Flahertys from Connemara – pirates, navigators and sea traders – they were mostly good to the islanders. Later the English took over and introduced a series of absentee proprietors who were not.

In 1922 the Land Commission did a fine

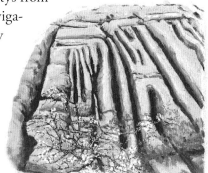

Limestone

job in buying out the island and selling to the tenants for long term payment.

Walking back we were charmed by the glow of spring flowers in roadside crevices in the carboniferous limestone of which the islands are made. Dr. Tom Molyneux told me how the limestone was deeply buried under pressure until erosion brought it to the surface. Then, through release from tension, cracks appeared. The limestone contains no more than 2 per cent of insoluble matter so when it dissolves in water very little material is left to form soil. But a little humus and plenty of rain make conditions beloved of plants like Alpine Gentians, Hoary Rock Rose and Mossy Saxifrages.

Next day we sailed south through Gregory Sound, then west along eight miles of two hundred foot cliffs, illuminated by the low sun and framed by long shadows.

They drop sheer into the sea with overhangs in places; a dozen or more layers of limestone showing golden yellow, blue, grey and green. Layers are separated by parallel horizontal cracks or fissures. Two hours sailing along the drama of these multi-layer precipices is enough to leave a brain-haze for months. I can picture them vividly as I write. The picture by Ros gives a better impression than any prose description.

A visit at Easte,r 2004 showed the Arans unchanged in the things that matter. Still there the soft sound of spoken Irish and genuine welcomes. There are still four busy sidecars. Mini-bus transport is duller but quicker and safer and the wit and wisdom of the drivers is as twinkling as ever. Currachs, upturned and resembling huge black slugs, still occupy stands beside landing places, but not a trading hooker is to be seen.

Heating is by oil, so turf is no longer sailed out, nor is seaweed exported.

The picturesque red petticoats and cloaks for ladies have gone. The sea birds, the grandeur of the cliffs, the arresting pattern of stone walls remain. There are airstrips on each island, a choice of good hotels, restaurants and shops.

Most pleasurable of all is the clean invigorating air, the brilliant light and the silence. When a gale blows on Aran, houses shudder and the swell thunders in the caves, but when the wind stops the ineffable silence is as near to the embodiment of peace as one is likely to find in this life.

The Arans are unique among islands. They have become a world-wide tourist cliché but one, like the pyramids of Egypt, that does not disappoint. Dun Aengus Heritage Centre told us of 200,000 satisfied visitors a year.

You can't go too often – there is always something new to see.

Inishmaan - The Middle Island

*'The Great Gaels of Ireland
Are the men that God made mad.
For all their wars are merry
And all their songs are sad.'*

Chesterton

As we turn back east along the awe inspiring Inishmore cliffs let's hear about how a passage was made in days when currachs were the only means of transport.

Here is the way J.M. Synge, greatest of all the Aran writers, described a crossing from Inishmore to Inishmaan about the year 1900. Picture him lying in the bow of the currach, with his shock of black hair and Hitler moustache wet with spray, facing the backs of three rowers, each with a pair of narrow sculls. The stern was occupied by another passenger.

"Out in the sound the currach rolled and vaulted in a way not easy to describe at one moment as we went down into the furrow, green waves arched and curled over me. I was flung into the air and could look down on the heads of the rowers as if we were sitting on a ladder or out across a forest of white crests to the black cliffs of Inishmaan. The men seemed excited and uneasy and I thought for a moment that we were likely to be swamped. In a little while however I realised the capacity of the currach to raise its head among the waves and the motion became strangely exhilarating. Even, I thought, if we were dropped into the blue chasm of the waves, this death with the fresh sea salt in one's teeth, would be better than most deaths one is likely to meet."

As we will see a lot of currachs on following visits some description may be useful here.

The sea-going Irish currach can claim a unique system of construction and design little changed for two thousand years. Its invention was a major development of 6th century Irishmen from the even older circular coracle. The primeval requirements of a craft that could be made without the use of forge or sawmill and where timber was scarce, dictated design in the west of Ireland up until a few years ago. Materials have changed as tarred canvas, and now fibre glass, have replaced greased hide for the covering and iron fastenings have taken the place of thongs. But the essential proportions for safe and minimum effort rowing, the need of a high bow for launching into surf and of lightness to allow for a quick snatch clear of the water, remain the same. To any one who enjoys small boat handling an

Currachs

unladen currach is a particular joy and the very thinness of the skin between ribs inches apart gives one feeling of being in touch with the works of the Lord and his wonders in the deep in a way never quite equalled in a planked boat.

It is not surprising that 'canoe' is what currachs are called on some parts of the coast.

In 2004 with John, a bird expert from Strangford, we crossed to Inishmaan on a sea almost as rough as that described by Synge. Running before a westerly gale at 15 knots in the sleek white ferry *Maid of Aran* had been exhilarating. As we watched gannets diving and stormy petrels walking on the waves, the *Maid* waggled her bottom and spread her snowy wedding train fifty yards astern, just like any lady enjoying herself.

We first looked at St. Columba's Chapel, the only one in the Arans named for this most famous of seafaring saints. It was roofless but retained its mystique.

Walking north from the pier we next saw a clutch of boulders known as 'Dermot and Grainne's Bed'. This comfortless spot recalls the greatest of Irish love stories. Dermot O'Dyna was the Galahad of the elite fighting men of Ulster known as The Fianna. Handsome of course, honourable most times, skilled fighter always, he was a great draw for the ladies – 'kind in hall and fierce in fray'.

Dermot owed fealty to Finn, Captain of the Fianna, so found himself in a quandary when Finn's second wife Grainne, daughter of Cormac, High King of Erin, persuaded him with potions to elope. Finn's fury knew no bounds and he sent the Fianna to capture the couple, dead or alive.

Cormac's trackers pursued so closely that Dermot and Grainne had to move each day to keep ahead, day after day week after week. There are many versions of what happened next. My favourite account came from James Dixon, the Tory Island artist.

> *"Them boys were on no hurry to catch up, for Dermot could 'a killed the lock with a swipe of his sword. He loved her so much that the heather under them was scorched everywhere they lay. Sand was put down by the trackers to mark each place and in it they planted heather. So where you see the two together be sure it's a Dermot and Grainne Bed. The best of it is that any girl that gets a courting on one of them is sure to have good luck. There's a big one up on Balor's Fort".*

The tale usually ended with a wink from Jimmy, a bachelor, to any young lady listener. Years later Dermot was tricked into hunting deer on Benbulben, above Sligo Harbour, armed with only his light spear. There he was killed by an enormous wild boar, the one now on the badge of Clan Campbell. With his big spear Dermot could have easily killed his assailant.

Grainne, broken hearted, was returned to Finn. After a year of moping she ignored the advice now given to railway passengers, stuck her head sideways from a chariot in motion and was killed by impact with a stone wall.

Dun Conor

animal or insect but a fruitful hunting ground for beachcombers.

Ros set off solo to explore south west. John and I went east to look for birds.

What did we see? Stones of course and wagtails, robins, blackbirds, jackdaws, more stones, oystercatchers, terns, black-backed gulls, jenny wrens, lots of primroses and baby briars in crevices. Most delightful were the 'chilloops' of a family of red-billed choughs playing overhead, and puffins doing up-tail dives for fish.

The wind had dropped and the mantle of the sea was scintillating silver below the sun and dull pewter to the north. Dun Conor's majestic skyline crest seemed to scan our every move, as it has scanned tens of thousands of visitors over more than two millennia.

We found its walls enclosing an oval area as big a football stadium. Nothing but fragments of small stone huts called clochans were to be seen inside.

These igloo shaped dwellings, commonly known as beehives, are made of concentric rows of flat stones, successive layers overlapping inwards over the one below. The upper surface of each stone slopes outwards so that rain is drained off and air for ventilation gets in. If well-built, as in the Skelligs, they outlast thatched or slated buildings by hundreds of years.

Maan is in two halves – the northern half is a plateau of limestone ending in broad beaches. South of it, a steep scarp rises two hundred feet. On the face of the scarp the pub, most of the houses, shops and businesses are to be found. On top is the great stone fort of Dun Conor.

The rear or backside of the scarp slopes for nearly two miles to sea level. This area has good grazing and massive 'storm beaches'. This is the surveyor's name for a rampart of boulders, piled up by onshore breakers and worn smooth by the friction of countless gales. They are inhospitable for

Dun Conor's walls, 20 feet high and 18 feet thick, are easy to walk along. From them, the view stretches 60 miles from Loop Head on the Shannon to the south, Slyne in the west and Errisbeg framed by the Twelve Pins up north. This castle does not frown; it reposes with an air of assured grandeur.

This is hardly a fort in the sense of a base to fight from, more a stronghold into which an entire population and their stock could flee until help arrived. Food and water would be a problem but privation for a week or two was preferable to death or brutal enslavement.

On the sister islands east and west are traces of multiple civilisations – from wedge tombs to sprung mattresses; Iron Age forts to concrete piers – manual signalling towers to electronic image transference. Herein lies the charm of travel in the Arans, the changing layers of culture in close juxtaposition.

Returning, we looked at Kate O'Flaherty's cottage where Synge stayed while writing *The Playboy of The Western World*. Ros catches the snug homeliness nicely in her pastel. There he dried his soaking clothes while the cricket sang by the glowing turf, listened to many a tale and became beloved for his sympathetic descriptions of the islanders.

We met for lunch at the long bar of the thatched inn which has figured in so many accounts, and Ros described as Synge's Seat where he meditated daily, rain or fine, looking across to the cliffs and forts of Inishmore. The conversation beside us was mostly in Irish among crowds of youngsters here to polish up their pronunciation and have fun while doing so.

Evening in the Ostan was deliciously relaxed – a fine dinner sitting in armchairs in front of an open fire with Bernadette, the proprietor, in close attendance as well as keeping an eye on the Dining Room which was in use for a gig.

In the Knitting Factory next morning we saw an original collection of linen and woollen scarves, jumpers and cloaks. A mail order service takes Maan garments all over the world - an admirable example of what a well-managed island enterprise can do. While Ros shopped and John, a linen spinner by trade, discussed the supply of yarns for knitting, I picked up some local history in the airy showroom.

To those not in a hurry, the island offers a timeless tranquillity.

J.M. Synge's Cottage, Inishmaan

Inishoirr

Spelled Inisheer on older maps, Oirr is a couple of miles in diameter, shaped like a slightly elongated circle. It rises 200 feet and is crowned – surprise, surprise – by a round fort of stone. Clare's frowning coastline is just five miles east, so Oirr is the closest of the Arans to Ireland.

Richard from Killarney, Stephen from Donegal and I waded ashore there from our own currach in 1965. At the back of the beach was a rusty standpipe supplying half the houses in the village. From it we drank excellent water. Rowing is thirsty work and the day had been hard, pulling close under the cliffs from Loop Head. No one took any notice of our arrival. Comings by currach were the norm.

A fleet of them were busy ferrying out goods and stock to be hoisted perilously into a steamer. Later our Dingle currach, longer, slimmer and more harmoniously curved than the Oirr ones, was admired by onlookers. Each island tends to have its own design.

Inishoirr Lighthouse

On, or should I say in, a sand hill at the end of the beach we

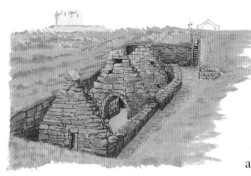

St. Kevin's Church

admired St. Kevin's Church. It is a replica of the famous one at Glendalough and the drifting sand here has acted as a perfect preservative. The excavated walls can be seen from above in attractive detail.

On another visit with Ros and John Andrews, 35 years later, we landed at the new concrete pier. The ground where we'd previously disembarked was built over. But the expansion has been well-planned. The close-packed blue slate roofs, many with dormer windows, remind me in the pleasantest way of similar villages in Brittanny.

We visited a charming Holy Well at the end of the western road. It was blessed by St. Ennion who I suppose is the same man as Enda of Inishmore, the profusion of Saints on the Arans can be confusing. The well is sunk in a little enclosure of rocks, topped by a triangular stone with three holes in it, said to be a fertility symbol of special significance for girls wanting babies.

Stone walls form each side of every road. They tend to be higher than those on Inishmore, six feet in many places and just one stone thick with boulders at the bottom bigger than most men could lift. Lots of holes let the wind through and give the beasts inside a view which is said to stop them straying.

The walls form essential windshields and in their lee grow bonsai-size white thorn as well as flowers.

Finally we came to the magnificent fort on a green hilltop. The approaches are steep and encrusted with field walls. In the midst of the pre-historic stone circle is a medieval O'Brien square keep. It soars 60 feet high with accommodation up top, arched store and guardrooms below. The keep in the Chevalier's Chateau at Tremazen in Finisterre looks just the same, a mark of sea links and exchange of architects.

Formhall was built by an O'Brien, fifth in descent from Brian Boru. The family liked coming here from inland forts in County Clare, on the excuse of inspecting the garrison, to enjoy sea air and fishing, but this ceased when the fort's south east corner was blown out, probably by Commonwealth soldiers, and it became uninhabitable.

Below us the wreck of the steel coaster *Plessy* showed the power of the sea. A huge wave put her ashore almost intact above high water during a fog in 1990.

Inishoirr Fields and walls

The Plessy

She has become famous as a background to episodes in the popular *Father Ted* TV series. Her cargo had included 'a few cases of whiskey' which were of course all taken over by the Garda – except, we were told, for a few bottles cached in a sea cave. "A fella from Galway that bought her made over a hundred thousand", said our guide of this highly visible wreck.

After a morning of rock hopping it was time for lunch. The Strand House combines accommodation with the Post Office, general store and outside tables at which to eat sandwiches.

The owner was full of local news. Oirr, he told us is best of all the Arans. It has the boldest currach men, winning footballers and original creative art.

To aid the latter there is an airy centre for carving, boat

building, painting and writing. All the Oirr houses are owned by islanders : they want to keep it that way and much more …

At the pier we saw boxes being unloaded marked, *'Wild Salmon – Caught in Galway'* to be smoked here for onward sale.

A dealer on the ferry confirmed that Oirr cattle, reared out of doors, make tasty beef. He had bought two dozen to be shipped next week.

Smallest can be best in islands as well as elsewhere. Island pride is like family or regimental pride. If a unit doesn't feel that it is the best, it is often the worst.

After a unforgettable day who are we dispute Oirr's claims of supremacy?

An Aran Tail Piece

If asked to judge I'd give Oirr the church prize for the sunken chapel, Inishmore the fort prize for Dun Aengus and the seamanship prize for its lifeboat crews. The welcome prize would go to Maan for the kindness we received in the Ostan.

Each of the three Arans has its own character; they are as different as dogs, cats and monkeys.

St. Macdaras and its Neighbours

Isles for all seasons

As we sailed west after a different visit to Inishoirr, the chart showed astern of us St. Brendans Isle, the only one called after the famous sailor.

North east was Hare Island, once used to hold unfortunate hares for coursing in Galway city. The Leverets, for a pair of banks beside it, is a neat bit of recent nomenclature. Mutton Island off Galway City was once grazed by sheep but is now inhabited by an underground sewage plant.

Along western parts of Connemara lie more isles, like sprawling Lettermullan, Crappagh, Furnace, The Skirds, Golam with its signal tower, Lettermore, Gorumna, Mweenish, Masson with an ancient church, and a legion of others. Animal and bird names get preference – Duck, Horse, Sheep, Rabbit, Crow, Chicken, Turbot (because of its shape) and conical Deer Island which used to carry an Irish Red stag and his hinds.

In early morning we slipped round Mac Point at the west of Macdara's Island and anchored off the conspicuous church.

This humped granite isle, seven hundred yards long, lies a mile off Mace Harbour. Its importance comes from association with a young churchman. His parent's choice of name had been unfortunate – *Sinach* meaning a fox. This word, like hare and rabbit, pronounced on shipboard brings enough bad luck to make a fishing boat give up for the day. So the coastal people called him Macdara, Son of the Oak, after his father.

His ability to extract a fish-hook painlessly from an ear and to tell where to net salmon or bring in shoals of herring, lead to his status as Patron Saint with an island of his own.

About 550 AD he cleared away some traces of Bronze Age activity and caused an Oratory to be built. It is almost square, twenty feet long by seventeen wide with walls two feet thick and a distinctive high pitched roof.

The isle soon became an 'inviolable sanctuary'. This meant that a person after committing a felony could not be arrested there. By staying for a year and a day – no small feat with winter waves breaking clean across – a pardon could be secured.

But don't try it nowadays!

A wooden effigy of St. Macdara 'stood for many ages' within the oratory until Malachias Queleus, Archbishop of Tuam, caused it to be buried 'for weighty reasons'. The mind boggles!

The only inhabitants today are a few sheep but on 16th July, the Saint's Festival, large crowds arrive at Mace Harbour. Boatmen do a fair trade and the pilgrims walking round the Penitential Stations have fun as well as a spiritual lift. These Stations, often marked on Admiralty charts, are unique to Irish sacred islands. The Celtic Church in Ireland was one of the first to arrange pilgrimages, both local ones round a revered place and others much further afield.

Macdara gets three stars in the Wallace and Ros calendar of islets!

Get there and have a walk around if you can.

Gull & Oystercatcher Eggs

Old House on Inishlacken

Inishlackan

'Little gem of all but islands'.

Paene. paene insularum
O venusta Sirmio'

The poet Ovid so described his favourite peninsula, as 'almost an island' on Lake la Garda two thousand years ago – the romantic love of islands goes back a long way!

Inishlackan is only just an island. It must at various times in the past been joined to the mainland. A shallow sound, a quarter of a mile wide is nowadays all that separates it from the shore. Nevertheless, Lackan feels 'offshore' and has a strong individual character.

Ros and I were taken out in a currach by Noel after he had sold his morning's catch of prawns on the pier at Roundstone. The voyage is about a mile and a half.

Ahead we could see on the peak fifty feet up a square con-crete water tank – hideous but useful as a landmark in the fog which is not uncommon in summer. Before its erection water was got by bucket from diminutive shore-line wells, only available when the tide was out.

The population reached 500 in the 19th century. By about 1965 it had reduced to a point where it had to be abandoned, leaving behind almost a hundred dwellings, now mostly bare-walled homesteads.

In 2004 the island felt deserted as Ros and I landed in

Looking north at Twelve Pins, Inishlacken

the neat wee harbour at the northeast end. We walked half a mile south along a straight lane. In places it had walls almost five feet high, big as those on the Arans, to our right lay the rocky hillocks and rushy bogs of the interior. A few cottages have been re-roofed and looked snug. The stone work endures with fine flat boulders used for jambs and lintels.

Bodkin House is a rather grand name for one which marks the tenancy of the family that used to own most of Omey. Mountain View Lodge lived right up to its name that October morning with the Twelve Pins gleaming magnificently down from the north under an ever changing pattern of clouds. The matchless brilliance of Connemara skies seem to draw additional tone and luminosity from the inland loughs below.

By the ruins of Travally village we watched flocks of sanderling and ringed plover feeding on sand-hoppers. Insects abound in the plum-coloured piles of rotting seaweed along the high water mark.

A short thick-set figure seemed to move on the skyline. I thought that we were seeing things. But it turned out to be a real islander with a spade. Paddy had spent most of his working life on 'the Lump' in London and Birmingham. Now he lives on the mainland but rows across regularly to fish and tend his veggies.

The path narrows there before curving on past The Goat Rock and the end of the island.

There is a Holy Well. One or more of these are to be found on almost every island hereabouts. They were probably held in respect long before Christian days. The veneration of wells by the Celtic Church has been attributed by some experts as stemming from the Desert Fathers living in Sahara sands where wells were as rare as they were essential. Crossing over to the west side we rose two snipe where a lough had become covered in rushes.

Another nest of houses sits by the west bay of Duirling Mhor. Gun Rock is marked on the map at the outer end of the north west reef. A six pounder might have stood there once to guard the approach to Roundstone but would have been very exposed. At the north end a shell midden is comprised of relics of feasting in pre-history occupation and perhaps desperate measures in recent famines.

Back at the harbour with half an hour to wait for Noel's return we had our picnic on a beach in the sun, entertained by sheep being reluctantly mustered on the pier. The ram made a sporting attempt to mount a couple of ewes but they were too closely packed for pleasure. In disgust he leapt off the pier and made his escape. That tup knew the game and could have breast-stroked for Ireland. The remainder were handed down one by one for passage to the mainland in a currach. It was an ageless, almost biblical, scene.

Lackan had given us the feeling of a much loved island, about to come back to life as more mainlanders move out.

Island Shepherds, Inishlacken

Slyne Head Islands

Our next objective was to double Slyne Head. It is a bug-bear that has held up sailing vessels on coastal voyages for weeks at a time in the past.

Slyne, like Barra Head in Scotland, is not a headland at all but the extremity of a chain of nineteen islands. This makes it a happy hunting ground for those who love secluded anchorages among the rocks. Such nooks here must be approached with great caution as tidal streams are strong and swells run high. There are several ways through the chain, each valued as a way to avoid the tide rips outside, which are terrifying in strong weather.

Twin lighthouse towers sit on Illaunamid, the outermost one, with a landing for stores on its inland side.

Chapel Island, forty feet high and rocky, was summer home to a hermit. Its fragmentary ruins make it the most interesting of the chain. Duck Island, the largest of the group, has a pleasant anchorage on its north side. Good for lunch but not to linger. Seas in a gale break right across the half mile long island.

A lobster fisherman described Cromwell Sound, the widest pass, through clenched teeth as *'a dirrty bit of water named after a dirrty man'*.

Inevitably the chain includes a stronghold. This is Doonavaul at the inshore end with some traces of a fort. From there, dues were levied by O'Flaherty kings on vessels using Joyce's Sound Pass, the most common way through.

Slyne Head marks a division in the sea as well the land. South of it the water begins to resemble the lighter blue of Brittany and florescence becomes more brilliant at night as the temperature gradually rises.

But we were bound north now and glad to see ahead of us the dark bulk of High Island. First we planned to visit Omey of the extensive shallows and blowing sand.

Omey Island

Parent of High Island Abbey

From our anchorage off the pier at Augrus it was short row by dinghy to the west end of the island.

Compared to its offshore neighbours, Omey is big. A square mile in area, it is shaped like the head of a boar with its mouth open facing west. It is tidal, accessible by car or foot over magnificent firm, clean sands a mile long, for the bottom half of the tide – that means for most of the time at neaps.

A chequered history has given it strength of character.

The strand packed with cockles and sand eels ensured early occupation by man, some of whom, in later times, were Druids who sound like romantic old chaps in white night-shirts until you read about their penchant for mass human sacrifices.

By the 5th century AD Christianity overcame them. Imaid, to use the old name for Omey, had at its west end a

Cruagh Island from Omey

Sands on Omey Island

monastery founded by St. Feichin of Fore. This bold fellow also set up the original Abbey on High Island. Omey, at its peak, schooled three hundred novices at a time and the resident monks continued for hundreds of years to look after pilgrims on their way to High Island.

Sadly all the walls ,except for one gable, are now covered in sand.

Edible snails, introduced from France, which can be seen round Omey wells are a pleasant relic of this occupation.

Omey's secular population peaked at 500 in the 19th century and seems to have suffered from several bad landlords. Among them were 'Soupers' who offered food and shelter as bait for conversion to their Anglican Church.

The island's Catholic Church, once the centre of a large parish including Clifden and Cleggan was almost sanded over until excavated a few years back. Better go soon if you want to see it, as it looks like being sanded over again.

In a sandy nick just above the western shoreline by Tra Rabhach, my cousin Mo showed us a Sacred Well that was revered long before the coming of Christianity. It was neatly kept in 2005 and a mere trickle, but the water tasted good.

Omey is a gorgeous island for a walk, specially on a breezy day with the Atlantic sparkling.

Omey Races on the sands at low water are being re-instated and should become as big an event as in days of old. A Meeting is vividly described in T. H. Mason's classic, *The Islands of Ireland*.

High Island

'I see but cannot reach the height
That lies forever in the light'

Longfellow.

High Island is indeed high. It rises two hundred feet on a base fifteen hundred yards long by six hundred wide. That makes it about a twentieth of the size of its neighbour Inishbofin. In shape it resembles a bat heading north with wings half extended. Viewed from sea level *Ard Oilean*, to use the Irish name, appears a huge lump of sea-shattered black rock. We had passed it often but each time on a sea too rough for a landing. Would we make it now?

I eyed Admiralty Chart, Number 2708 – it lists 'Abbey in Ruins', 'Brian Boru's Well', 'Copper Mine Shaft, and three 'Penitential Stations'.

The usual place to land is where the island narrows to a fifty yard neck near the north end. A long cliff-bound gut provides a little shelter from the swell.

Cascades of white water poured off rocks as we approached. But fortune was on our side. Feichin Mulkerrin, the owner, was there in his currach. A skilled seaman with knowledge of many islands, he is named after the founder of the Abbey and has lived in sight of *Ard Oilean* all his life. The right hand side of the gut is a seventy degree slope of gritty schist. Barnacles give its rough surface some additional stickitude. A weathered rope hung down like a donkey's tail from a rusty stake. But as we approached at low water it was almost out of reach. Getting ashore didn't look like a cakewalk, so with customary gallantry, I said, 'Ladies first!'.

Ros made what the Galway Blazers would have called a 'tremendjous lep'. She hit well up the rock, slithered half way down to the water but managed to grip the rope. "Come on," she said, "its easy!" Feichin held the gunwale hard in to the rock. I got finger tips on the rope and scrabbled for a toehold until my feet found a tiny ledge.

Only when you've scrambled up on all fours do you realise that there is lots of grass up top. The sheep grazed there were famed for the sweetness of their mutton. But shepherding is hard and currently the only grazing is in winter by barnacle geese. As we got our breath at the top a heron flapped slowly by.

Said Feichin, "It's only a couple of miles by air from Aughrus".

A heron's only enemy would be the peregrine falcons of which we saw no sign. In the past an 'eyrie' of young hawks was an annual tribute to the King of Connaught. A low stone building is known as the Miner's Hut after a gang who dug here for copper in 1829 for 'Clemency Dick' Martin of Ballynahinch. A deep vertical shaft yielded little and I wonder if the workers were ever paid. They appear to have done major damage to the Abbey, removing stones to find eggs, birds or valuable grave goods.

Pollen counts have shown that High Island was cultivated in both the Bronze and Iron Ages. Each community in turn suffered from the difficulty that only in the most set-

tled weather could any boat be left afloat. That meant that between voyages it had to be light enough to be hauled up a fifty foot cliff.

The Abbey we'd come to see was founded by Saint Feichin, a Sligo man still much revered in Connemara. One account describes him as *A man of bright summery life, fair spoken, an abbot and anchorite'*.

Picture him as short, raw-boned from fasting, ribs scarred from lying on stones for a bed, with deep-set kindly eyes that could light up when he spoke of faith. He knew when to encourage weaker brethren and when to be rough with lay-abouts. Once he threw his crozier into the millpond below the Abbey causing water to flow and drown a millwright sleeping when he should have been felloeing the waterwheel.

St. Feichin died of the yellow fever in 664 AD.

A sheep track led up through an airy pass between grass banks. I followed it, still pinching myself to be sure I really was on High Island after so many disappointments. Halfway down the far side Ros discovered the well named after Brian Boru, Emperor of Ireland.

Everyone's favourite Irish hero, he played the major part in driving the Danes out of Dublin. This was no mean feat when other Danes under Canute were becoming masters of all England. Brian learned from The Norse the importance of sea power, captured some of their galleys and used them as models to build his own. So, on a day when there was a favourable wind, Brian ran thirty miles from his family castle on Aran to visit High Island. St. Gormgall

North East landing, High Island

the Abbot was his soul father. With a name meaning 'the blue eyed stranger' he was probably of Nordic blood – tall and erect with a long flaxen beard, clad in a white woollen 'leine' under a linen stole. Now Ros and I were standing on the very spot where that great shaggy king drank from the well. He must have glanced at the tumbled grandeur of the Twelve Pins to the east, Slyne Head to the south, Shark and Inishbofin to the north and the surf beating on the Sister Rocks below. I could almost feel him looking over our shoulders as we watched waves in a tide rip bouncing up like the flight of peewits above them. Of Brian an Irish Annalist wrote

'He was not stone in place of an egg.
He was not a wisp in place of a club.
He was a hero in place of a hero
And valour after valour'.

The Well is said to cure colic and similar complaints and a sip did a power of good for my holiday hangover.

Below us the land fell away in gracious folds to a sloping hollow, the only sheltered spot on the island. In it we could see the gables of the tiny Abbey, barely eight foot square inside and the rectangular stone wall close round it. This has recently been restored by the stonemasons of An Ducas Heritage Society. Enclosing both were traces of the 50 yard wide cashel. This was probably built as a sheep corral centuries before the monks arrived.

Slithering down the hill we picked our way among scattered rocks to enter the ruins.

Side by side outside the east wall of the Abbey were the graves of eight monks. Each lay in a coffin made of slabs of flat stone. The most elaborate, lined with white quartz, could be the resting place of Gormgall who The Annals record as dying on 5th August 1017. All eight graves have been dated to the same period, the most active in the Abbey's history. Gormgall must have had a busy life, for besides being Abbot and personal priest to the Royal Court, he was chief Confessor of all Ireland.

What fantastic sales his diaries would have made!

Just outside the east wall of the cashel is a large beehive hut with its roof intact. Archeologists have found traces of several churches nearby. We trod pathways worn smooth by the feet of inspirational saints – Columba, Patrick, Brendan and maybe Malo from the port named after him in Normandy. He was known as a jolly fellow for his love of islands, specially the Brittany ones. There are also traces of a scriptorum and of huts to house pilgrims. Engraved cross stones of distinctive design are found all over the island.

Details of these fill the 200 pages of *High Island* by Jenny White Marshall, uniquely beautiful among island books. When the faithful occupants of the eight graves were alive St. Feichin would have been remembered as a remote figure, as old fashioned as the eight now seem to us.

High Island monastery was to last 500 years. How many of today's institutions or communities can expected a life as long?

Monastic Site, High Island

Visiting sailors are best to leave the details to academics. Too close a study can spoil the sense of wonder at the bravery, skill and perseverance of the monkish builders and carvers. For centuries they lived in fear of death by drowning, starvation or murder at the hands of pirates. But they kept on creating and protecting.

Years ago I heard a story on Ile Molene, inside Ushant, of a monk from there going off to an Irish island. Now it was not hard to re-create his feelings.

He came here to advance his skill in illuminating manuscripts and found it a hard school. Day by day he could hear the sea crash against the rocks and the gulls cry as he watched the scarred face of the headlands. In a westerly gale the sea leapt fifty feet in the air and sometimes white water would come roaring down the valley in which the Abbey lay, threatening destruction to paints and parchment.

When he was decorating a document with whorls, patterns and intricate drawings, he forgot the privations. But in winter there was scarcely eight hours of daylight. Fine work by guttering candle was almost impossible. Boredom set in.

In spring a soft wind would blow from the west and Slyne Head made him 'think long' about its resemblance to Ushant.

At such times the faith of the man from Molene must have grown dim. Endless island afflictions depressed the soul – the rule of silence, no news, no snug bedclothes, never more than three hours sleep between prayers, short rations, frequent fasting and no girls. Worse still enduring the penance of a stone pillow for going outside the Abbey walls without permission. It was hard to fully accept the Abbot's teaching that privation and hardship was the only way to communion with God.

He thought of his apprenticeship in Concarneau where a monk could walk around the Ville Close, mix with the people and decorate neighbouring churches.

Each autumn a wine ship from Douarnenez called at High Island with supplies for communion, but each autumn he put off taking a passage. He'd miss his Irish friends and a runaway monk with the distinctive Irish tonsure could not be sure of welcome in Brittany. Next year he might go home.

There is no record of Vikings landings – but that doesn't say they didn't take place. Burnt churches leave ashy traces to last for centuries and none have been found.

How was High spared when so many suffered massacre? The monks may have been numerous enough to swing axes on cliff tops and defend their treasures.

The Abbey closed in the 12th century – perhaps because the great days of island monasteries was over but more likely because of inaccessibility. Bofin, Caher, Clare and others with easy landings carried on for another two or three centuries.

The dissolution of monasteries policy of King Henry VIII and his successors accounted for what was left.

High is an island of sacred soil and great distinction, 'the most extensive remains of an early mediaeval monastic community in Galway'. It has the only specific record of an island visit by a High King of Ireland.

Seals on Friar Island

Friar Island, beside High, has a sheltered port between stacks at the east but mysteriously no traces of religious occupation.

High has had six owners in last 150 years, perhaps an example of the French saying that an island gives you pleasure twice – once when you buy it and once when you sell it. Its integrity is safe in the hands of Feichin though at some stage it would be good if a Trust could take over.

If you are lucky enough to get there, treat it with the care and respect it deserves.

A Tail Piece

While Ros was far off exploring, I had sat down alone on a seapink tussock with a Robinson Crusoey feeling when a wheatear (my favourite shore bird) perched on a stone beside me, turning his delicate pink breast to the sun.

"How do you like it here?" I asked before putting on a squeaky voice for his imagined reply.

"Great" he said, "I arrive at High Island about the middle of May with a bunch of my Whinchat relations. Soon find myself a pretty wife. No traps, snakes or weasels and not too many humans. Peace and quiet! Lots of stone wall for nest sites so its easy to rear a family, sometimes two!

We can't stand the cold so come September we're off again – three days to the Straights of Gibraltar if the Portuguese Trade Winds are blowing, then a sunny winter in Africa!"

He gave a flick of the white bottom after which he is called (from the good old Saxon word *arse*) and flew up to bag a daddy-long-legs and carry it across the lough for lunch, I felt quite envious of his lifestyle!

Pirates and Sea Raiders

'The O'Malleys are much feared everywhere by sea.'
Calendar of State Papers; 1599.

Preamble

The following pages describe Clare first because it was Granuaile's main base, then Inishbofin, Inishturk, Achill, Achillbeg, Inishbiggle and those in Clew Bay. All were used at times for her fighting ships.

The mid 16th century was the most dramatic period in a thousand years of island history. Columbus, within living memory, had discovered America. The development of the three-masted ship which could beat to windward and keep a cargo dry below decks was shifting the balance of power from the Mediterranean to western Europe and Iberian caravels sailing round the Cape of Good Hope were breaking the Arab hold on the spice trade.

Sea-borne trade from the continent to the Arans and ports like Galway and Sligo was increasing. The Spanish made four attempts during the century to land troops in Ireland for trade and to support their Catholic co-religionists. Queen Elizabeth neither wanted nor could afford war in Ireland, but with the risk of invasion it was essential for her to control the whole country, specially the larger islands and coast. The Clans were under threat as never before and desperately needed leadership.

The other seagoing tribes competing with O'Malleys for ships to board were the O'Flahertys, who commanded the approaches to Galway, and the MacSweeneys, known for their courage and striking features, who looked on Donegal seas as their own.

And there were more as a contemporary quotation shows:

'The English at sea in the guise of brigands bring merchandise too and treat every ship they meet as an enemy without distinguishing whether it belongs to friend or foe. This evil is accentuated by the Dutch acting in same way.'
Contarini History in Quotations, Cassell 2004

It was an age of piracy!

Onto this stage, with perfect timing, stepped Grace O'Malley with a force of character that comes only once in a century.

Granuaile on Clare Island

'Storms, adventure,
Heat and cold
Galleons, currachs,
Buried gold'.

Anon.

Arriving at Clare Island is always exciting, specially under sail. It is artistically sited, with mountains on either hand. Three miles long by two wide and rising 1500 feet, it has an appropriate silhouette of a lion couchant.

Written as 'Clere' on 17th century manuscripts, it has given its name, as well as shelter, to fascinating Clew Bay.

In early June 2004, the 30 foot ketch *Agivey* came in from the west after a fight with white horses in the tide race off Achill Head.

That dusting made the shelter provided by the cliffs of Knockmore doubly welcome. "Salmon cages, ten of 'em, Begow!" was a surprised cry from the lookout off Port Lea on the north shore. It seemed an exposed position but they have survived for some years. As we rounded Kinacorra Head, the peaceful harbour we'd been expecting wasn't there. Instead was a gridlock of barges, ferries, tenders and currachs in a sweat of activity. Local lobster boats and fish farm tenders lay at moorings in zig-zag rows. A floating crane was driving piles for the new deep water pier. All was

go-go to be finished for the holiday season. Tourism was taking charge.

For all the lack of room, we were guided in alongside the inner pier and care taken to see our warps would not be crushed by passing dumpers. The island people are true seamen, and as helpful as ever.

It was here Granuaile lived in summer. A most pleasant place and, above all, strategically placed at the centre of what became her sea kingdom.

In the 16th century Clan O'Malley had been policing the western seas for generations, exacting fees from passing busses, carracks and luggers; sometimes providing pilots to skippers who did not know the coast. Fishing here was good enough to draw crews from afar. Deciding which nations could fish, and where, needed a good head and strong arm. Another service of hiring galleys to Hebridean chieftains for transporting barelegged Scots, known as Redshanks, on raids was a riskier but possibly more profitable activity,

Granuaile was born about 1530. She acquired the name Granuaile from the Irish *Grainne ni Mhaille*, mispronounced because the 'Mh' sounds like 'W' as Grainne Mhaille. She grew up to be galley captain, sea trader, sea raider, virago, and diplomat – above all a leader who broke all the rules except the ones that mattered.

In a sea fight picture her with pistols and sword at waist and musket in hand.

Ashore afterwards she would be swearing and spitting, swigging firewater and gambling with gusto. For other ends she could don a silken dress in the latest Spanish fashion

Sailing to Clare Island

The O'Malley Castle, Clare Island

and play the part of an elegant court lady of charm and poise. Thus she managed to bargain woman to woman with Queen Elizabeth in London and flirt with her Deputy, Sir Philip Sidney in Galway. In all the wide west she was the most famous personality. Men would follow a woman like her to hell and back.

She made trading voyages to Spain. The passage took six days if the wind was fair, offshore across the Bay of Biscay to Galicia. Tough seafaring that was, exposed to wind, spray and strong light for twenty four hours a day.

Her first marriage was a political one to Donal, son of the Clan O'Flaherty chief. His forbears had sacked Rathlin, Country Antrim, as early as 735 and continued to take their toll of Galway traffic. In this company she polished the art of boarding ships. Cargoes passing the Bunowen coastal castle included spices, silk, wine, wool and gold. A plea for help from Galway Council to Dublin Castle in 1553 gives the feeling:–

> *'O'Malleys and O'Flaherties with their gallies … have been taking sundry ships bound for this poor town, which they have not only rifled to the utter overthrow of their own-*
> *ers and merchants but have wickedly murdered divers of young men to the great terror of such as would willingly traffic …'*

Granuaile, aged 23, soon took control of the O'Flaherty fighting and trading ships.

Intelligence of Spanish plans was one thing beyond price

to Queen Elizabeth. Were they going to send an Armada up the Channel or try a left hook by using Ireland as a bridge to invade England? When were they coming? How many ships? Any galleys included? No one could provide better information than Granuaile. While selling her hides and dried fish in Ferrol or Corunna she could listen to the dockside gossip and see if ships were fitting out for war. Warnings, if urgent, could be got to London within days. Perhaps she did just this, and by doing so saved her neck on a couple of occasions when in a Government prison.

When Donal died fighting ashore, Granuaile married politically again. This time to 'Iron Dick' of the Bourke or Lower MacWilliam family, another land fighter. Through him, she was able to garrison Bourke castles north and east, a strategy which worked but only through many a shuffle in alliances. For Granuaile, life without adventure would have been intolerable.

Many myths grew up around her. She gave birth at sea to a son, later called Tioboid na Long *[of the Ships]* who would inherit both his mother's diplomatic and seafaring skills. An hour after the birth her galley was boarded by Turkish pirates. The battle on deck was almost lost until she appeared wrapped in a blanket and shot the enemy captain. Her crew rallied, captured the pirate vessel and hanged the crew. If this did happen, it was at a time when Turkish galleys under Kheyr-ed-din Barbarossa and his successors held the whole Mediterranean in terror. The year 1570 was recorded as a peak for the slave markets in Algiers. Demand

was incessant. So Turks and Sallee Rovers had every reason to explore Irish waters for loot and infidels to carry off.

After their lateen-rigged xebec that met the O'Malleys disappeared without trace, it was most of a century before more appeared in Irish waters.

Granuaile was always for the quick riposte as demonstrated by a call at Howth on the way back from imprisonment in London about 1576. She was refused entry to the Castle on the excuse that the Lord was at dinner. Spotting a child playing nearby she identified him as the heir, tempted him on board and carried him off to Clare Island. He was returned safe after the Lord had promised always to leave the gate open at dinner time and an empty place laid at the table – a practice the St. Lawrence family carry out to this day. But don't try sitting in it!

In 1596 MacNeil of Barra in the Outer Hebrides infuriated Granuaile by killing some of her clansmen. She retaliated by taking her galleys direct 300 miles to Barra, storming a MacNeil castle and returned laden with loot. It is difficult to convey in a few words to someone who has not spent days at sea in a crowded open boat the stress that two-way voyage, with a fight in between, must have involved. It was exhausting, but the stuff of life for 66 year old Granuaile. Risks judged aright and plans fulfilled – the look in the eyes of men she'd led to victory.

In 1603 after her death, Tioboid was confirmed in his estates and created the First Lord Mayo! I wonder if she would have approved.

Granuaile spread her assets. She kept captured ships in

the natural harbour at Inishbofin, or inshore close to her castle of Carrigahowley, (properly *Charraig a Chablaigh*, The Rock of the Fleet, anglice Rockfleet) in Clew Bay. There the mooring ropes of her galley were said to be tied to her bedpost. Alternatives were Doona on Blacksod, Kildavnet on Achill or Louisburgh near Roonagh.

A signalling system linked the fleet. A lookout on Knockmore, Clare's 1500 foot peak, could send or receive orders. These would be repeated on intermediate stations like Inishturk or Achill to the farther out ones like Doona.

One smoke column on Inishbofin could mean, 'Intercept ship coming north'. Four columns might say, 'Enemy fleet in sight. Stand by to repel attack.'

Flags of different colours and sizes gave other information.

Sligo-bound ships could be intercepted by galleys in Portacloy. Those based on Clare Island gave a much better chance of a fair wind than others who had to start from inshore.

Granuaile's Castle by the harbour today has walls intact but a dank floorless interior and nil atmosphere. From there

East Window, Clare Abbey

she could be at the tiller of an armed galley and off in chase within minutes.

Granuaile was wild as a sea eagle, a gallant figure who pranced when others plodded and whose courage has acted as an inspiration to western sailors, including myself, ever since. A dozen biographies have been published but there remains enough mystique to attract more writers with the sea in their blood.

Clare was the jewel in the crown, her stronghold and lookout.

Of special interest is an Abbey founded by Cistercians in 1224, the most striking building on all our western coasts. Open to the winds for centuries it has recently been handsomely re-roofed. Traces of mediaeval frescos, including hunting scenes, can still be viewed on the ceiling and walls.

The flora and fauna of Clare Island have been more thoroughly tabulated than on any other British or Irish isle. The first survey, completed in 1911, under the auspices of Dr. Praeger and the Royal Irish Academy, identified 3,500 plants species, (585 new to Ireland), and 5,200 species of fauna (1,200 new to Ireland). A number were new to science. The fact that so many species could inhabit such a small area came as a revelation to the academic world.

Ninety years later, what's more, a follow up survey was undertaken. The Royal Irish Academy published it in three paperback volumes in 1999. The number of plants has decreased, but the bird species have grown. A pair of gannets now nest on an inaccessible stack at the west. Choughs, the most attractive of the crow family, flourish and peregrines have an eyrie on the cliffs.

The Surveys reveal the cruel hardship experienced during several 19th century famines. The population peaked at 1,600 in 1841, then halved by 1851 and reduced to 500 by 1870. By 1890 the O'Malley owner was bankrupt through his efforts to remit rents. Then, the Congested Districts Board bought the island for about £5,000. They successfully re-organised land holdings to equalise the quality of each and sold out to tenants on a long term repayment. The population is now about one hundred.

Clare is a beautiful isle of many faces and many moods – well worth getting to know.

It seems a little short on legends and early artefacts. Perhaps all the investigations have chased the wee folk away but they'll be back before long.

Caher Island

'The most beautiful thing we can experience is the mysterious'

Einstein

As we sailed west from Clare Island the grey wedge on our bow grew gradually larger. It seemed at times to stand on a white pedestal of foam that told of high swells and scrambly landings.

Rolling gently south we had time to scan a 6 inch OS map in addition to the Admiralty Chart. It showed three sites named after St. Patrick – a circular stone caher or fort at the south end, a bed and a church within a circular cashel – all beside Port-a-Temple, the landing place on the east. I cannot think of another island with so many Patrician attributions. St. Columba was there too.

Clare fishermen had told us that Caher was bracketed with Inishglora and Macdara's as the three holiest islands in Connaught. Sailors traditionally dipped their topsails in passing each and sent up a 'wee mouthful of prayer'.

Looking at the swell, I silently did the same.

Caher is regarded as St. Patrick's farthest west excursion and I see no reason to doubt that he spent quite a lot of time there during his 27 year mission. Indeed he would surely have been glad of respite there from his missionary work.

On the mainland he was often at risk of assassination by Druids, who remained powerful at court. Other courtiers

Caher Cross

were jealous of his success, but on Caher he was revered and beloved.

It may have been an O'Flaherty who gave Caher to the church and later O'Malleys gave protection which was much needed. Clonfert as an example, founded by St. Brendan on the Shannon, was plundered nine times between 650 and 1150. Six raids by local tribes, only three by Vikings.

Granuaile came here once to capture a McMahon man who had murdered a Spanish lover of hers on Achill Island.

She hanged him with her own hand! Her motto was like the American Joe Kennedy's – 'Don't get mad. Get even!'

At first there would have been a solitary hermit on Caher. Such men lived out the tradition of early Desert Fathers. There were no sandy deserts in Ireland but the sea made a more than adequate substitute. Later, like-minded friends put up bothies at a suitable distance and joined the eremite. A community grew, an Abbot appointed and his flock increased.

Perhaps as a result of St. Patrick's sojurn, the monks of Caher Island wore a largish size in hats, (if they ever put on such a thing) but curiously no name of a Caher abbot has survived.

Caher is about three quarters of a mile long with 200 foot cliffs on the north west tapering to low reefs at the south east. Once it was covered in scrub but today it doesn't grow enough timber to make a wooden leg. Our slow approach was a time to try to conjure up the scene when the island was a hive of monkish activity with currachs curved like black bananas floating lightly on the water, their crews fishing for mackerel and pollock. Beside them seals doing the same. Oaken boats towing timber out from the mainland, others landing wicker baskets of scallops for food and grain from Inishturk. Perhaps a boat deep laden from High Island with mutton or copper from its mine.

Life in a Celtic monastery was harsh; food was sparse and fasts were held each Wednesday and Saturday. Prayer started in the dark at 2.30 am and continued at three hour intervals – Terce, Sext, and Nones – throughout the day. Monks who

Agivey off Port-a-Temple on Caher (Clare & Achill in the background)

failed to observe the rules could be severely punished by being put on half rations or sent on exposed pilgrimages.

Nine o'clock Vespers was often sung and the sound of monks voices and harp music would waft euphoniously across the water.

St. Columba and St. Brendan could have chosen Caher to discuss Atlantic islands and the wind patterns needed to reach them, to pray for monks in danger or foundering in mid-ocean, in the knowledge that so perishing they were doubly blessed and divine reward assured. They would have talked too of the need for a central authority to weld together the ever growing number of monasteries.

Double luck came as we arrived. Not only was the swell low but Joe O'Toole, a friend from a previous visit, arrived from Inishturk to collect a ram.

'I'll show yous round', he offered as his currach put black tar marks all over our white topsides.

A handsome stone cross of great antiquity stood at the head of the creek. Others showed in the background.

Joe enjoyed being a tour guide. 'There were two or three hundred monks at a time here', old folks had told him. 'They called it Cathair na Phadraig or Patrick's Caher. There's St. Patrick's Bed just outside the church door.'

A flat slaty slab met our gaze.

'He must'a been a wee man for it's less than five foot long and too narrow to curl up on. But one thing's sure – any sick person that sleeps on yon' gets better quick'.

The stones on the altar, we were told, are all holy and most have stories of special powers and where they arrived from. I was careful not to touch the pumice one that could raise a violent storm if moved.

We are all part of the past. Caher makes you feel it.

Inishbofin

Entering Ireland's finest island harbour for the first time is a nervous business. There are leading marks to pick up – a white tower on a rock on the east side of the entrance, another on the far off shore behind. They can be hard enough to see on a misty day.

If you let your boat get off the line to port the Bishop and other sunken rocks wait to grab your keel.

Think how much more difficult it would have about 1660 facing the plunging fire of a dozen guns and forty muskets at point blank range from the Cromwellian Castle. If you survived the missiles there was the possibility of a sunken chain across the narrows to set your ship aback.

A gentleman called Bosco had earlier occupied the fort site. He was a huge figure, 'meadowed chin to navel in an acre of black man hair', as Squire O'Halloran described to me – a pirate, Spanish perhaps, who held sway for around fifty years after 1600. That was just after Granuaile's time. 'The Squire' was the island patriarch in 1952 with a fine head of white hair and rare turn of phrase.

He told of an enemy ship chasing Bosco who saw a way of escape in the narrow gap between Bofin and Lyon into shal-

Entrance to Inishbofin Harbour and Bosco's Castle

St. Colman's Abbey

have been packed gunwale to gunwale.

The fort is expertly designed – clinging close to the sea rocks, square with the corners drawn out in turrets to give defensive fire to the curtain walls and mounting 22 guns.

It was held by Royalists against Parliament in the Civil War but taken over by the Cromwellian ships in 1652. A garrison was maintained there against the Dutch for 50 years thereafter. Reveille, other bugle calls and the tramp of booted feet must have grown familiar to islanders.

As we peered into a massive well in the courtyard a pair of red-billed choughs dived close overhead. Their nest, low down in a chimney, was full of hungry youngsters.

Once inside the harbour on lawful occasions you are secure from any normal gale.

The pleasant time-warp between our first arrival and one in 2005 was Mrs Margaret Day. In 1952 she dried our clothes when we were sodden youngsters and gave us fresh vegeta-

lows where he could not be followed. Close hauled and just weathering the Bofin shore he warned his crew,
"Don't cough!"
This was lest it upset the perfect trim of his sails …
Absence of details of Bosco's deeds is an infuriating gap in Bofin history.
At peak fishery times in the 19th century, the harbour is said to have held 150 fishing schooners. If so they must

bles. In 2005, bright and witty as ever, she plied us with coffee and fresh baked scones and lots of island stories.

Father O'Malley on our first visit sent a parishioner to get gas cylinders we needed from Letterfrack telling the messenger that they were 'for three young Protestants'.

From him we heard that the earliest Christian arrival was that of St. Colman, perhaps the most famous Irish Saint of his day. As Bishop of Lindisfarne in Northumbria he was outvoted at the Synod of Whitby in 664 on the vexed question of the date of Easter. Departing with thirty supporters he decided to found a monastery on the Isle of the White Cow (the translation of Inishbofin). He picked it for its key position where coast-wise traffic would be bound to call. His charming abbey, as depicted by Ros, was sited by a small lough in a sheltered glen a mile east of the harbour. The Celtic monks had a great sense of humour and Colman would have been amused to find his name linked 1500 years on with mustard.

Granuaile used the island to good effect, the only place she could keep her fleet afloat in winter. She had a fort at the north entrance to the harbour and another promontory look out at the south end.

Bofin is an island of exceptional beauty with winding lanes, heights and hollows by the port, a sheltered beach to the east and rock stacks to the north. In spring my favourite birds, whinchats and wheatears, are tame and common.

By its harbour is a brand new 32 bedroom hotel run by Margaret Day's sons and grandsons. There is a choice of other accommodation, pubs and restaurants as well as excellent connections by ferry to Cleggan.

East Town Inishbofin

Inishturk

'O, its a snug little island!
A right little, tight little island'

Thomas Dibdin, 1640

Inishturk is way-out but in a central position, around five sea miles each from Clare Island, Inishbofin and the mainland. It seems to have mostly kept out of the strife which has as so often brought bloodshed to the coast. One massacre is recorded when Bingham, son of the English Governor, came out from Galway at a time when his family were at peace with the O'Malleys. Having stayed the night he signalled to his bodyguard who murdered his host and hostess and all the islanders. A similar story is recorded of Omey Island in Connemara so maybe it all happened there.

Approaching Turk (as the island is colloquially known) from the north it looks high, with humps at either end and a knobbly ridge between. As you get nearer, the village shows up white in the lee of the northern hill, then the harbour mouth can be discerned at the left hand end. From the south it looks like a big cat.

When we anchored *Caru* there in 1952 the island had no cars, a hundred horses, a single exposed pier and lots of currachs. No ferry ran to Ireland, just a weekly link from Inishbofin. A shebeen near the harbour offered refreshments to thirsty sailors. A fine stone wall divided private land from the commonage which amounts to nearly three quarters of the island. Passing through it we explored the 400 foot western and southern cliffs. Off them are stacks with musical names like Alnarehoo, Boughil Mor, Boughil Beg, Turlinmore, Carrickboola and Glassillaun. Between these, ooghs or 'oweys' penetrate the headlands. Most of these rocky guts are seal-haunted and have caves at their inner ends.

On the south east we scouted Portadoon between Dun Ballyheer and Dun Ballycrugga. This pretty, natural harbour is a hundred yards long with an entrance too narrow for anything but a currach.

We were asked into a cottage overlooking it for a cup of tea and boiled eggs with Mrs O'Toole Peter. The sobriquet is important because most of the population are O'Tooles who came here from Wicklow.

About 1960 I stayed with Derek Hill, the artist of Tory Island, at the home of Mrs O'Toole Peter while he planned paintings of the unique western cliffs. On the day we were to leave for urgent affairs ashore, the sea was up. All looked hopeless. It was the only time I have felt trapped on an island … then suddenly through the murk appeared the bluff bows of the old wooden Inishbofin mail boat with Pat Concannon in the wheelhouse.

Soon in his good hands we were safe in Cleggan. It was those hands which had saved a currach crew in the awful disaster of October 28th, 1927. Bofin boats had made good catches when an icy blast struck them at dusk. It was too strong to row against. Other craft from all along the coast were in desperate trouble. Some made the expensive decision

Green Island, Inishturk

Mary Jo's Chickens on Inishturk

to cut their nets and get into shelter. Many crews drowned when their currachs were blown onto rocks and broke to pieces. Others were caught side on by big breakers, filled and sank. For nine long hours Pat kept his currach head to sea by hauling on a frail line attached to his nets so as to crest each rising wave and ease her into the succeeding trough.

By the morning he was a hero. Despite fingers chaffed to the bone he had saved both his crew and the nets vital to their livelihood on a night which claimed the lives of nine Bofin men and thirty from other ports.

Arriving in *Agivey* in 2004 we found that Inishturk had moved on greatly. It still has no rabbits or rats but lots of cars and a charm all of its own. Near the harbour are comfortable boarding houses and a District Nurse with a custom-built Health Centre. Beside it is a new building office for the Community Development Manager. Affairs are run by a non-profit-making Co-op which owns the Post Office, a shop and the Club which has the only liquor licence. Water from a well-dammed lough is piped to all houses. The main road is tarmac and steps have been built down to several beaches. The Prendergast family build and repair currachs. Four smartly painted tributes to their craft-manship lay on the slip. A regatta is held about mid July to which visiting yachts are welcome.

For communications there is an illuminated helipad and several ferries a week to Roonagh. A pier, built in 2005, pro-vides berths for fishing boats and ferries at all states of the tide though the harbour is still a neat little box which recalls the wee ports of the isles off western Norway.

The fishermen were helpful, vacating a berth where we could lie, propped against the inner buttress at low water. Their gift of crab claws was much appreciated.

Here in the full fresh fragrant morning it was the dawn chorus that awoke us; and all day the song of blackbird and thrush echoed off the cliff at the south side. The water at high tide showed a rare shade of bluey green.

Paddy O'Toole, who is writing a book about the links with St. Columba, interrupted his house building to come and chat in our cockpit. A burly man with face and hands used to out of door work, he is also an academic and told us how analysis of pollen in deep soil samples indicates two distinct breaks in human occupation – one Dark Ages, one Mediaeval. These are fascinating clues to island history, so look out for his book!

In the evening we had to climb six hundred feet to the Club. But excellent draft Guinness made it worth the ascent. We had the good company of Danny the Development Officer who described how Turk is full of enterprise and character. Its population remains stable at around 80, rising to 120 when the school children are home on holiday. We wished we could have stayed longer.

Achill Island

A Dugort schoolboy is said to have written that Ireland is a small island off the coast of Achill. I hope he got extra marks!

Achill is distinguished as the largest and loftiest of any Irish island. In shape it is like a pistol with the barrel pointing seaward – an ancient battered pistol as the shoreline has many indentations. The surface varies from golden beaches and bird-haunted saltings to steep bare mountains. The size is twelve miles east-west and ten north-south.

It projects from the coast in the same way as Skye does from the West Highlands in Scotland. The resemblance does not end there. Both islands are now linked by bridge to the shore. Both alternate between huge flat expanses of heather and soaring mountains. Both have magnificent western cliffs. If you arrive across the bridge and veer off towards Dugort the landscape at once recalls the grandeur of Skye.

Slievemore (2200 feet) rears in isolation on the north while Croaghaun (2192 feet) and Mweelaun (1500 feet) form Achill's mighty western headland where a herd of wild goats survive. A pair of spikes called The Ears of Achill also survive as this ridge remained un-glaciated when projections on the rest of the island were worn smooth by the passage of 2000 feet of ice. Today the Ears are a leading mark for sailors dodging Pluddany Rocks inside Inishkea.

It is these features along with top quality surfing, sea-bathing, fishing and hill walking that draw tens of thousands of visitors in summer and thousands to build the holiday homes which now proliferate round every corner.

The name Achill probably means eagle, a reminder of the efforts being made to re-introduce these lordly birds which used to be commonplace breeders on the sea cliffs.

Granuaile, in charge of the O'Malley fleet, saw how Achill's great bays could act as a trap into which to drive a passing vessel if it refused to obey the signal to heave to. Arrests at sea must have been frequent – Captain Cuellar a survivor of the Armada wrote, 'Scarcely a day passes without a call to arms'.

Imagine the stirring sight of a flotilla of galleys speeding out of Inishbofin to round up a pair of Spanish wine ships in 1570 – capture one intact and drive the other onto the Achill rocks.

Armada wrecks provided rich pickings in 1588 but discretion was the word. Authorities in Dublin with no more than 1500 soldiers available to guard the whole country were rightly concerned that ships' crews might, if boldly led, establish a stronghold in the west. Helping Spanish survi-

Deserted Village, Achill Island

vors was declared punishable by death. For years after 1588 individuals proven to have done so were tried and executed. So we hear nothing of Granuaile's activities at this time.

But she was able to help hide away a few individuals when the thousand ton *El Gran Grin*, one of the largest in the Armada with crew of 300, was wrecked on Beetle Head at the south west point of Clare Island. No names, no pack drill.

At Achill Head vertical blasts whizz down off the cliff top while tide races produce perpendicular waves below. Twin hazards that could hold up a fleet for weeks by the need to round it – and while waiting could run short of food or be attacked by land.

For friendly vessels of moderate draft there was a passage inshore of Achill at high water. Clan O'Malley secured the south entrance to this by building Kildavnet Castle. The north was controlled by manning the old fort overlooking the channel known as the Bull's Mouth between Achill and Inish Biggle. Turbulent tides there run up to eight knots and make an angry clatter along stony shores but inside is good shelter for ships to anchor.

At the north end of Blacksod Bay, Granuaile's own galleys could be dragged overland at Belmullet and so reach Broadhaven and the north coast, however rough the sea was, to gain the weather gauge on enemy ships.

When we took the *Aileach* through the re-opened Belmullet Canal in 1991 an old man told me, unprompted, that she did that regularly. Keem Bay (pronounced Kim)

offered a useful outer anchorage where galleys could lie in ambush.

So Achill gave the O'Malleys an exclusive 40 mile inshore passage from *Cahir na Mart* (as Westport was known in Granuaile's days) to Broadhaven and Donegal Bay.

Our arrival on Achill in *Wild Goose* in 1975 was at the small artificial harbour at Keel. Its entrance is sheltered by Inishgalloon where sheep can safely graze.

It was here in the fifties that sunfish or basking sharks were landed after being trapped in nets and speared to death to be cut up for their oil – a cruel and bloody end for the most peaceful large fish in the sea. If undisturbed these creatures, up to thirty feet long, cruise slowly along just under the surface gulping plankton through huge open mouths, but if attacked they 'gave you the tail.' One stroke was enough to destroy a currach and drown its crew. So bravery and skill was needed to catch them. The slaughter was justified fifty years ago when sunfish were plentiful and the means of livelihood scanty.

My friend, Hans Ott, an ex-German Army Sergeant, at one time looked after the commercial side at Dugort. Joe Sweeney of Achill, who befriended us when we were trying to get the swing bridge joining Achill to the mainland into working order to pass through, was a major operator. The season lasted 16 weeks and employed up to a hundred men. The sale of oil from livers covered costs – several barrels could come out of one big fish and *The Mayo News* of September 1951 notes the price as £80 to £95 a tonne.

At that period 1600 sharks were taken one season but such

Achillbeg from Achill

a number was exceptional. An outlet for the flesh to make feeding stuff, not easy to find in such remote area, was needed to make a profit.

After a few years the number of sharks diminished and the Achill men, like Gavin Maxwell of otter fame in Skye, found it was no longer worthwhile. Can't say I'm sorry. The stench of unburied bodies was most unpleasant.

Whatever its drawbacks, the fishing created a hundred good jobs at a time when the alternative was kelp. It took 25 tonnes of wet seaweed to produce a tonne of kelp to sell to plastic factories at no more than £5 – 'a terrible slavish kind of a racket', as an islander put it to me.

Recently tourism has steadily increased to provide less odorous employment.

From Keel Harbour you can see on the south side of Slievemore the famous Deserted Village. Almost a hundred stone houses – roofless but mostly well built – are preserved as a tourist attraction.

Looking six miles south you can see The Bills, three isolated rock stacks resembling a mini-Tory Island. They have fine caves and arches and great fishing but not enough grass to be worth grazing. Galleys would have found them useful to hide behind before making an interception.

Achill has attractions a-plenty but if you enjoy cliff scenery try and see them by boat from seaward. When gilded by a golden dawn Para Handy, the Scottish coaster hand, would have described the scene as 'Chust sublime.'

Achillbeg

This is a two-humped island, dwarfed by it large neighbour.

Curiously it has more traces of Christianity than on all of Achill, also on its west side the largest promontory fort on the coast. Within this is the base of what may have been a Round Tower, the most effective refuge for monks when attacked. With the only access by a narrow door 15 feet up, assault was near impossible. Raiders rarely had time to stage a prolonged siege and after a few days would look for easier meat.

On the east is a lovely beach off which we anchored waiting for the strong tide to slacken. This gave time for a bathe and a scenic walk on its west side to the lighthouse, no longer lit and the abandoned village containing a few attractive holiday houses and no shops or services.

The Sound between Achill and Beg almost dries at low water. A boat trip from Darby Point takes less than ten minutes.

Inish Biggle with ferry in foreground

Inish Biggle

'Said Inishoo to Inish Biggle
'We can always have a giggle.
Davillaun girls are full of sport
But their island has no port.
Inishkea still smells of whales,
The sea breaks right across in gales
Here we sit in waveless shelter
Even if it blows a welter.'

W. Clark

Biggle is of curious shape appearing on the map like a stage horse looking east.

In the lee of Ridge Point are narrows less than two hundred yards wide between Achill and Inish Biggle They are known as 'The Bull's Mouth' from the way the sea roars when an eight knot tide runs against wind.

About a mile wide and low lying with just one hundred-foot hill, it is a good place for a walk. We passed the Post Office and Church, then the Mission House to which families came a century ago at the instance of a Mr Nangle who tried, by offering food and accommodation, to convert Achill islanders from the old faith to Protestantism.

Climbing Mount Biggle we perched on granite boulders to look down on Inishoo and the entrancing pattern of islands. Beside us the marram grass rippled like water before the wind, the sunshine marking each stem in something between silver and white, the colour of light. Wind had carved the dunes into curves and crests as if it was snow.

On Annagh, Biggle's elongated sister isle, Inishkea people used to build sod huts in summer. This was known as boo-leying when boys and girls, free of parental control, watched the flocks and a good time was had by all.

The Biggle Ferry in 2005 was a 20 foot currach, as it has always been. Under oars it could take a long time and was usually a bit of an adventure. Now with a powerful out-board engine and life rings on the foredeck you will get safe across in minutes. Slack water, when the tide eases off, is the best time to go.

If conditions are not right, be prepared like The Spartans to 'sit on the sea-wet rocks and carefully comb your hair.'

Clew Bay - The Bay of Islands

'There is a tide in the affairs of men which taken at the
flood leads on to fortune.'

Shakespeare

Clew Bay is rectangular, six sea miles from north to south, twelve the other way. The inner part contains some eighty inhabitable islands. If you include tussocks and bits of salting the number is over three hundred.

The isles are drumlins of limestone scoured by ice sheets with a cocktail of gravel on top. They come in all shapes

Clew Bay

– ovals, lozenges, snakes, dragons, arrowheads or long-legged beasties.

Only about six are dwelt on all year round in 2004 but holiday houses are on the increase.

For Neolithic people 'The Bay', as it is fondly referred to by those who live near it, was a paradise of sheltered waters packed with fish, a separate isle for each family and easy transport to market. Today, to sail among them is irresistible to boat lovers. They see the isles over shimmering water as a blue undulating barrier. Their upper edges are hemispheres, half-moons, or twin

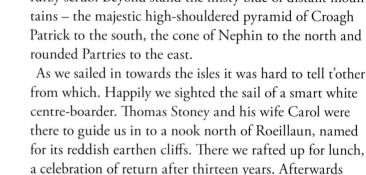

Croagh Patrick

humps; some have outlines like cheese slices, others eroded to rounded topknots as in Ros's painting.

The colour combination is of azure and silver sea, a dark band of foreshore and above it slopes of emerald grass or furzy scrub. Beyond stand the misty blue of distant mountains – the majestic high-shouldered pyramid of Croagh Patrick to the south, the cone of Nephin to the north and rounded Partries to the east.

As we sailed in towards the isles it was hard to tell t'other from which. Happily we sighted the sail of a smart white centre-boarder. Thomas Stoney and his wife Carol were there to guide us in to a nook north of Roeillaun, named for its reddish earthen cliffs. There we rafted up for lunch, a celebration of return after thirteen years. Afterwards they piloted us to their home bay past Inisherky, Inishcooa, Inishitra, Inishdooney, Inishkee, Muckinish, and lots of others. The top of Rockfleet, Granuaile's tower could be glimpsed to port. Off each island there was at least one shoal, sandbank, or half-tide rock. Fish or mussel farms, not all well marked, were an additional hazard.

Soon we were snug alongside the grass-topped stone pier fifty yards from a sun-warmed blue door at the most delightful of elderly family houses.

Tradition here is that no Stoney is allowed to go sailing single-handed under the age of three! It was in this bay in 1991, with enormous Stoney help, that we fitted out *Aileach*, a replica 16 oar highland galley built in Donegal for a successful voyage to Stornoway.

Next day Thomas kindly lent us his boat, just the right size for the Bay. In it we first visited Islandmore, known for having the largest population in a village south of the 'terbert' (tarbert) joining it at low water to Knockycahillaun. Here the Gill family are traditional pilots for Newport.

We inspected Inishoo, an outer isle, popular for its wide beaches. Carol told us to look out for the White Calf. This creature of exceptional beauty was born on the island. In due course the family decided to bring it ashore. A party of six attempted to surround the heifer but only succeeded in driving her over the top of the hill. The far side was steeply sloping sand – but not a single hoof mark could be found, there or anywhere else. The owners, 'badly shook', sold the island saying it was the work of the wee folk.

Inishgowla (island of the fork) was the next call; its wide arms shelter an anchorage popular with fishing boats. Sheep ran away as usual at our arrival and eyed us from slopes that were sprinkled with red and yellow trefoil. The remains of four houses could be seen in the lee of the hill recalling a visit in 1952 to Moynish More where several families then lived. They supplied passing boats with the necessities of life like eggs, mutton, porter and locally distilled poteen. In pouring rain we were greeted by a stooped lady whose wrinkled face split into a wide smile. She wore the traditional patterned shawl, voluminous black skirt and red petticoat. Her feet were bare and I envied the way she could walk over the sharp pebbles.

"Ah, the craytureens!" was her first remark, as she saw my teenage sister and her friend Val. "Come yous all in and get warm."

In the dark smoky kitchen we watched her take a teapot off a high shelf and empty it of a ten shilling note, various keepsakes and a colony of spiders. Then it was hooked onto the crane over the turf fire. The tea leaves were roasted for an age before being wet. Soda bread and boiled eggs tasted good, as our sodden woollens steamed dry.

"My son works in England," she remarked, "You must know him".

Clew Bay 'Topknot'

A hard one to answer tactfully. We were on our best behaviour in receipt of the finest Irish courtesy. These islanders were (as Mason noted in his classic *Islands of Ireland*) *'remnants of an ancient race and gentlefolk in the best meaning of the word.'* We emerged blinking in the strong light to pay our respects to the Gentry Tree on the hill where the fairies dance at full moon.

In 2004 after leaving Gowla we carefully dodged Carraigeen-na-Francach (The Rock of the Frenchman), a bouldery reef crowded with seals. The name comes from a tragedy after the anchor cable of a Breton fishing vessel parted in a hurricane. She broke up on the rocks and her entire company perished.

It is hard to believe in summer calm how, during winter gales, white-crested breakers come roaring into every sound.

Then we climbed stiff-legged out on the pier by the Mayo Sailing Club at Rosmoney. Off this forty fine vessels lie on moorings. The welcome and help given to us was just as warm as when we were there ten years earlier. Our eyes were drawn by red dinghy sails to Collanmore Island, which gives shelter from the west. On it is an offshoot of the famous Sailing School from Iles de Glenans in Brittany.

Going west we watched a new pier being built on Inishlyre, near the homes of Tom and Joe Gibbons. The brothers with keen eyes, weathered faces and neat moustaches are skilled boatmen. Their family have been community leaders, ferry captains and harbour pilots for generations. Lyre was once a popular anchorage for trading schooners in pre-engine days. With ease of access from the sea past a lighthouse

on Inishgort, ships could discharge cargo into lighters and avoid the tedium of a tow four miles to Westport.

Clynish, farmed by two families, was proud of its new pier. There a few years previously I'd met Mary Hughes, 'Queen of the Bay'. The name came from her skill at finding big catches for anglers. She was busy 'casthraatin' lambs.' The team worked with such speed that I kept well back. Dressed in an Aran sweater there seemed a risk of being grabbed by a helper, pushed forward and treated before anyone noticed I wasn't a woolly. By 2004 Mary was based in Corraun by Achill taking deep sea anglers out west. Great people, Clew islanders!

Another day's sail took us to fertile Inishraher (anciently Rathnt, Ferny Island) near the south end of the bay. Ninety feet high, a quarter of a mile long and partly wooded, it once had a village of nine houses. Some are now being refurbished.

Its southerly neighbour Inisheany (*Aonach*, 'Meeting Place') lies on a main route to Westport so a natural site for trade between the mainland graziers and passing ships. The Annals of the Four Masters records in 1239 that 56 ships gathered on its shores. This was the greatest galley fleet ever seen in the west, captained by the O'Dowda of north Connaught. He was demonstrating against paying tribute to the O'Connor, King of Connaught.

The isle saw blood spilt about 20 years later in the first major western raid by Normans. O'Connor ships were in Clew Bay to harass the 'foreigners' and landed foraging parties. Those on Inisheany were surprised and killed by mail-clad Normans, borne in boats supplied by the O'Flaherty. In the follow-up all the islands of *Cuan Modh*, as Clew Bay was then known, were swept clean of cattle – their only portable wealth.

The myriad islands were vital to Granuaile. Her castle at Carrigahowley was carefully sited, just tall enough to observe an approaching boat or signal to lookouts on Clare; but not so high as to attract the attention of an arrival. A captured ship drawing up to 15 feet could be towed in close to the castle and allowed to dry out on a mud bank for leisurely looting.

After 1573, when Royal Navy ships like the *Handmaid* were permanently on the coast, Clare Island became risky as a base. There is no record of any galley, Irish or Scottish, successfully attacking an English warship or supply ship of the period in the open sea. A single cannon ball on the waterline would be enough to sink a lightly built galley. But should a warship venture to sail among the Clew islands she would be at acute risk from ambush before a gun could be trained. The same applied if attacking Belclare and Cahernamart (*Cathair na Mart* now Westport). Galleys could be dispersed and hidden up winding creeks.

'Clew is the most spectacular bay in Ireland,' Thomas Stoney had remarked and after three days of exploration we concurred. No other Irish archipelago has so many interesting islands surrounded by such strikingly beautiful mountains.

'Clew isles must be a bit 'samey'', said an American travel writer planning his trip. He was wrong. There is a family resemblance but on close acquaintance each has its character and most have intriguing stories as yet untold. Here we must end this brief description of lovely Clew Bay.

Sainted Isles and Soaring Stacks

Preamble

Can'st follow the track of the dolphin
Or tell where the sea swallows roam?
Where Leviathan taketh his pastime?
What ocean he calleth his own?
If you venture to answer such questions
And marvel at voyages of old
Then discover the Celtic Saints' seaways
In search for a land of God's own.

Apologies to Rowland Canymaid

The following pages describe twenty islands having an Abbey or Oratory of 5th or 6th century origin. This was a period rightly labelled the Golden Age of the Celtic Church. The men of God combined a strong simple faith with caring for all forms of life.

In far distant storm-beaten communities which seldom totalled in all more than a few thousand inmates, Irish monks achieved incredible distinction. This was in three diverse fields –

Firstly:- Their fantastic copies of biblical texts which lead Kenneth Clark in Civilisation to enthuse:

'The pure pages of ornament are almost the richest and most complicated abstract decoration ever produced, more sophisticated and refined than anything in Islamic art-hypnotic in effect. The clear rounded lettering of their Carolingian script carried the word of God to all the western world.'

Secondly:- With the above texts as tools Christianity itself was safeguarded by Irish monks when its faith and practice had been wiped out over most of Europe after the collapse of the Roman Empire. Men like Columba in Scotland and Columbanus in countries now called Germany and Italy, took a major part in re-establishing and spreading the Christian message. The texts, a thousand years before printing was invented, were an essential tool.

Thirdly:– Bold Irish navigators developed the sea-going currach from the circular river coracle and pioneered the exploration of the Atlantic. St. Brendan of Clonfert later known as The Navigator almost certainly got to north America by 600 AD five hundred years ahead of Lief Ericson of Denmark. Other Irish monks colonised Iceland by 670 AD, two hundred years ahead of the Vikings. Tim Severin, with whom I sailed in 1985 on the first part of his voyage in the leather currach *Brendan* across the Atlantic, told me of his worldwide researches into early voyages.

He remarked that he had found elsewhere no drive so strong as that felt by Irish monks in the 5th and 6th centuries to probe unknown seas. He specially admired their determination to discover, at whatever risk, lands where no man had been before.

They had the urge to worship God in isolation from the turmoil and strife of their day in Ireland. This is what lead them west *'peregrinare pro Christo'*. Many set out, of whom we have never heard, and perished in mid ocean. Others found their islands.

When writing the following pages I have had it much in mind that the monastic inmates included leaders of genius with worldwide vision and ability. These are men and women of whom Ireland can be justly proud.

Chief among them were St. Brendan as mentioned above, St. Columba who founded Iona in 563 and St. Cormac of the Sea who made many solo voyages that are barely recorded. Sometimes they went to sea rudderless and oarless, letting tides and winds take them to destinations known only to God.

The Inishkea Islands

Do you remember an isle, June?
Do you remember an isle?
Snug in the lee
Of green Inishkea
In a clear blue sea.
And the gift of a crab
With claws all a-dab
Ready to grab
You or me.

Apologies to Belloc

Inishkea North and South are the biggest of eight islands lying, a handy couple of miles, off the west side of the Mullet peninsula.

The hard Dalriadan gneiss of which they are formed gives them strength to make an ideal breakwater. My strongest recollection is the pleasure of the shelter they give from the Atlantic swell and an anchorage in water so clear that you wonder how a boat could be supported on such insubstantial stuff. Add an abundance of fish and it is not hard to see why calls here have been popular since boats were invented. And that was long before historians started writing.

The best option, if the tide is up, is to go alongside the pier on Inishkea South. Tucked behind a rocky point it has lasted with a little maintenance since it was built in 1888.

In 1952 when we first got there, the sea ran blue and

Anchored off Inishkea

laughing and the island looked like a slice of heaven. The village was a long row of serene cottages facing the white sand of the bay, with daisy-studded sward behind. The high green hill to our left looked down and said nothing.

Beside us men were handing lobsters from a currach into a wooden floating box they called a 'hotel'. They kept us a couple for lunch.

The skipper of a fishing boat, with a red, round, unshaven face, came over for a dram. Asked about the weather he was illuminating.

'We do be lookin' for a low from the Atlantic every wheen o' days. It's the same arl' the year round. A wee calm comes after each wan; maybe long or short. That's a mather of whether the low aims for Barra or The Skelligs. We count three bad days to one good in summer.' He took a long pull at his pipe. 'I'll houl' you this warm spell will break tomorrow for Friday goes agin' the week!' He was right!

On a call a few weeks later in bad weather the bay was empty and the island looked a dreary wind-swept waste.

Neolithic and Bronze Age occupation has left widespread traces here. In the 5th century monks arrived and built a church below the hill and some beehive huts. Only the bases remain. There was always a strong temptation to use the stones for ballast once the monks departed. We don't know when that was. A seven hundred year gap in island records ensues.

The islanders in the early nineteenth century revered a Godstone; it was shaped like a rugger ball and said to have been St. Columba's pillow. Possession of it was in contention between the two islands because its presence helped the crops. Eventually it was broken on the order of a priest.

'It was thrown into the sea, so it was,' said Frank Lavelle, last King of Inishkea North, my friend and informer.

In 1593 came the only recorded violence. Ulick Bourke and eighty of his clan took refuge on the islands, not then inhabited. Galleys to transport Irish soldiers to kill them were strangely supplied by Granuaile. We are told she was under the duress of Bingham, the harshest of Governors. Ulick escaped by boat. After a gallant stand on the hilltop, his men were all slain.

The islands appear uninhabited for the next century. After 1700 the population grew rapidly. It even increased during the famines of the 19th century as fishing provided at least subsistence feeding. Evictions are recorded by absentee landlords even in the worst famine of 1847. Food was supplemented by local piracy. Ships becalmed up to twenty miles offshore found themselves boarded by currachs. Hungry isles men drove the sailors below with stones and seized grain.

The ever ready Royal Marines then covertly manned the *Royal Victoria*. When she was boarded off Davillaun they came out of hiding and arrested the raiders. Protests by the whole population, and natural sympathy ashore, obtained release for most. But the piracy was checked.

The Inishkeans were inventive and industrious. As a linen weaver I was pleased when Frank told how flax was grown and yarn spun to make nets and garments, far more durable than those of cotton.

Ruined Cottage, Inishkea South

Several hundred tons of wheat and up to thirty tons of kelp were exported most years. The kelp kilns also produced lime from sea shells for mortar and spreading on the land. High quality poteen was another export and on offer in a shebeen on the island.

In 1908 The Congested District Board bought the islands and transferred ownership to the natives on long term repayment. Later the Board set up a fish curing station and arranged the building of wooden cutters big enough to make trawling possible and carry bigger cargoes than currachs. The Board also got merchants to stock ice so that catches could reach Dublin in good condition.

Cash became plentiful in 1908 when a whaling station was set up by a Norwegian company on Rusheen Island beside the south island jetty.

A pier for the catchers was built, also a slip up which 20 ton whales could be hauled by steam capstan. Two dams helped to conserve water and try pots to boil blubber were housed in high corrugated iron buildings.

The catchers averaged 50 whales a year – mostly Finn, but also Humpies and Sperm. Up to 60 jobs were split between Norwegians and south islanders. North islanders were excluded due to the jealousy to which most islands seem to have a tendency.

All the lobster fishermen made a killing as the offal brought their prey inshore. Money was good but the stench was not – 'bad enough to knock a man down at two miles distance'. Mainlanders were complaining at the same time as whales became scarcer.

In 1914 the fishery packed up. Whales nowadays are scarce. I've never seen one in Irish waters. Frank Lavelle told me that in the fifties he was used to seeing whales by his boat and found them gentle and much too clever to get into his nets. This was in contrast to the basking sharks which could do vast damage if accidentally entangled.

On 26th October 1927 several islanders saw a strange currach which left no shadow and refused to answer a hail. Two nights later thirty currachs went out to fish as usual at dusk. Just as the first had filled their nets a fierce blast of icy wind struck, far too strong to row against. Some currachs, maybe ten, which had stayed in the lee of islands fought their way into shelter. Others who made the bold decision to cut away their expensive gear early did the same. Some boats were swamped, others dashed on mainland rocks. Nine young men were drowned. The total along the coast was over 40, the worst loss of currachs since Brecan's fleet of 50 were swamped in Rathlin sound in 400 AD. Two more Inishkea men were drowned hunting seals in 1929.

A report followed recommending how the Relief Fund set up to help the bereaved should be distributed. It describes how,

'In one cottage said to be 200 years old an aged man and his sister had no bed and slept on the floor in straw, shared with a goat and hens. The chimney was just a hole in the roof.'

The losses and privation took the heart out of the island.

St Columba's Church, Inishkea North

Irises, Inishkea

Inhabitants often found themselves cut off for a month in winter and days in summer, mainly due to lack of a sheltered landing on the Mullet.

Both Inishkea communities, tempted by easier living and schooling on the mainland, passed a vote in 1934 to abandon the isles. Leaving an island is much more than moving house on shore. The solidity of the mainland stuns. Ex-islanders I have met seem to withdraw themselves in mental isolation. The pain is eased by returns to old homes each summer to tend flocks for as long as health and strength allow.

Frank commented that, "Too much money and too much education has the country beat." Then exulted, with a wide sweep of his arm that, "Here we had the finest fence in the world and all free – The Atlantic Ocean".

One moonlit night we rowed out from the anchorage and rested on our oars in the midst of the dark mirror of the sea and watched the island in a golden field. Inishkea looked like part of a timeless world, and so it is in microcosm.

Next day we tacked into Portahilly, a cove about 300 yards wide on Inishkea North. There we lay in water so clear that you could see every link of the chain three fathoms down, a-glitter on the sandy bottom.

The dinghy took us to a broad white beach. The pier here, built as relief work in 1863, only lasted two years. This irregular shaped island of 600 acres, rises no more than 30 feet apart from one conspicuous sandhill. It is in such a key position that it has had more than its share of calling vessels.

This was St. Columba's island. His little gem of a church,

built around 540 and the gables of half a dozen cottages lay to our west. To our east rose traces of an abbey and three mounds. These, referred to as baillies, are named *Beag, Mhor* and *Doighte*, 'small', 'big' and 'burnt'.

Who did the burning and when? One could make many a guess – Fomorian pirates, or Phoenicians explorers, Vikings, Norman freebooters or Scottish redshanks. They all came in for shelter on coastal passages. We climbed Bailly Mhor, 500 feet across at its base and 60 feet high.

Small sunken dwellings and stones with inscribed crosses can be seen on its sheltered side. In one chink a wheatear's nest of tawny roots and ready for eggs pleased me most.

Archaeologists have found traces of a dye works operating beside the hill – was it supplied by Tyrians bringing big purple murex shells or just the scarcer local equivalents? Other colours were obtained from dog whelks and local plants.

Islands need some undiscovered gaps in their past to preserve an air of mystique and here there is a full share. In the early 20th century there was a pub and small shop. After the island was abandoned, fishermen waiting to haul their pots would come in and sleep overnight in a house that still had a roof.

Sailing away we noticed at the north end the longest place name in Ireland – Carrickmoylenacurhoga – 21 letters for a 200 yard long half tide reef. The locals just call it The Carrick.

The islands in 2004 looked little different from 1960. The white of the cottages had faded to grey; a couple on the South Island had been tastefully re-roofed as holiday homes.

Faster fishing boats go to the mainland each night so the bays are lonelier.

The grass was less eaten down and birds tamer.

Frank's parting words rang in my ears, 'The best island I ever heard of to live on'.

Davillaun

This mile long island with a lovely name is rewarding for a couple of hours exploration on foot. In the galley *Aileach* we spent two uneasy nights at anchor in an exposed rock-bound cove on the south side. The landing there is a scramble over large boulders. There is an easier one at the east. Once on grassland it is easy going to the 180 foot summit. The row of houses there were abandoned about 1920 after complaints about excessive rent. Nearby are traces of an oratory and some interesting carved stones.

From the top you can see that Davillaun is the most southerly of the Inishkea group and lies off the tip of the Mullet. It has fine cliffs and stacks at the west where Leach's fork-tailed petrels have been found nesting. The light keepers from Black Rock three miles west had a narrow escape there when the currach bringing them to Blacksod got lost in a winter fog and they had to abandon ship and face a dangerous climb. Luckily no lives were lost.

Inishglora

Ahead was a low reef of granite and sand-scoured turf. This was Inishglora, holiest of islands. Hardly half a mile long by two hundred yards wide it has no anchorage, poor landings, limited grazing and no hill to shelter its inhabitants from winds that, 'would lift a cow off the island'.

In shape it resembles a mallet with it head to seaward; at the handle or inshore end is a cluster of buildings, now roofless, within a ruined cashel wall. These are the relics of a monastery founded by St. Brendan mentioned in the Preamble.

Why did the Saint pick Glora, poorest of the Inishkeas, for his church?

The answer is simple. In his age, physical hardship was looked on as the way to spiritual perfection, so the worse the island and the smaller the boat the nearer to heaven by mortification. Brendan was that brave explorer who, during a seven year voyage to many islands, got across the Atlantic and back. Another, Cormac the Navigator, is credited in Glora with three singled-handed attempts to reach a place where he could worship in peace. He was a pupil of St. Columba on Glora.

They'd tell you on the Mullet that it was in a cell on Inishglora that Brendan planned the voyage and probably made his last call before setting off 'in quest of many marvels overseas'.

My first visit in 1956 to Inishglora was on a calm sunny day. Its south shoreline where reefs provide a little shelter looks like sand but turned out to be ankle-breaking round boulders of assorted size. We anchored *Wild Goose* off the east end and were greeted by two men lying on the long narrow oars of their currach amid tangles of maiden hair weed. It was Frank Lavelle and his brother John who had a knack of turning up when wanted.

John wore a tattered Norfolk jacket and peaked cap with a bobble on top, left behind perhaps by some Edwardian goose hunter.

They offered to show us round.

Glora has the remains of several beehive houses. The largest, almost fifty feet in diameter now in ruins, contained three cells. Frank pointed out the one used by St. Columba. The place began to grow on me when Frank ushered us into the rectangular Oratory which judging by its thin walls once had a thatched roof.

Inside was a platform of boulders about five feet long by two wide. 'That's where St. Brendan slept', said Frank. Some of the top stones looked horrifyingly jagged. It became easier to picture the Saint lying there when we saw his one luxury – a bedside light. Begorra! It was a thin stone projecting from the wall, hollow on top for fish oil and a wick.

A gentle purring came from under the bed.

I groped into a hole to find out where it came from and a stormy petrel slid into my hand. I wouldn't have been more surprised if I'd found a leather satchel marked 'Brendan Sanctus, Ardfert Abbey, County Kerry'.

The wee bird lay fearlessly on my palm. This delicate

St. Brendan's Bed, Inishglora

creature could feed its chick on rich oil from fish fry and plankton, fly to America without touching land and find its way back to an Irish nesting burrow.

So St. Brendan was lulled to sleep by the purr of petrels nesting in the rocks below him as they do in the same rocks today.

I wouldn't be surprised if these tiny swallow-!ike birds had a way of telling him how to find fair winds on the bosom of the broad Atlantic where they spend months at a time on the wing.

Outside broken sculls and human bones, weathered to a sickly green, lay on the grass. The sand that covered them had blown away.

Frank showed us the Seven Stations of the Cross, some surrounded by sharp pebbles. 'Pilgrims have to crawl round them on bare knees', he explained.

Then the Women's House for pilgrims to rest in and the sheiling where the last family lived until 1940.

After that it was the graveyard;– 'No one living on the island was allowed to be buried there', he commented. When I fell for it and asked why, he added straight-faced, 'They have to be dead first, don't ye know'.

'But they still bring ones over from Cross on the Mullet where there used to be an abbey. The spirits of dead men cannot cross water so if there was a bad boy in the family, they liked to put him here to keep him out of the way'.

A Holy Well eight feet down worn stone steps was the next exhibit.

Asked about the meaning of the name Glora, sometimes spelled Glaura, Frank told us, 'It has nothing to do with glory at all, at all. It's the howl of the wind, the crash of waves breaking, and the brightness of the sea over the reefs that be's in it.'

That's a lot for one syllable but it was clear in the mind of Frank who knew his islands so well.

Then, as we came to the wee beach half way along the south side, he ran some grains through his fingers and told us about the anti-mouse sands;–

'T'is in use every day on the Mullet, for the sands of Glora are sharp entirely. No rat or mouse can survive here because their ears are open like sea shells and the sand in them gives a wild itch. A wise person would take some back home because it's a grand thing in any house.'

Then he explained the legendary side.

Glora was the scene of the Second Great Sorrow of the three in which the shanaghies seemed to revel.

Lir, the father of Mannanan, King of the Sea, had three sons, Aedh, Fiachra and Conn, also Finola, a daughter of matchless beauty. Eva, a jealous stepmother, put a spell on all four. They were condemned to spend three hundred years in the form of swans in the 'wild streams of the sea' around Glora. As a consolation they sang so sweetly that no music ever heard in Erin could compare.

In one version of the myth St. Kemoc turned them back to human form when the Queen of Connaught demanded to bring the swans home to sing at her dinner parties. In another it was St. Brendan who broke the spell. Either way there was a rub. The years of their lives had been so prolonged that they came back centuries old and expired right beside the Glora chapel. Frank showed us the mark said to be of a webbed foot made by the dying kick of the fair Finola. If you hear swans singing around Glora, specially if there are four of them, better keep to windward lest Eva happens to be around and you too be transformed!

But would it be so bad? Sometimes when having to return home to the problems of business I fancy it would be pleasant enough to spend endless time swanning among these lovely islands. If given a choice I'd rather be a stormy – it is the petrels who add Inishglora serendipity.

One of our crew lost her spectacles on that visit, so if in thirty years time you see a swan wearing a pair of horn rims or a stormy with an Irish Cruising Club bow tie it will be the *Wild Goose* crew doing time!

Frank gave us the princely gift of pair of lobsters before seeing us safely on our way. He was kind to us on lots of other visits to his home at Frenchport but on our last visit in 2004, God rest him, there was no Lavelle to greet us.

Glora still looked uncared for that year but if in future it sports tarmac tourist routes and teashops selling plastic swan's feathers I don't think I'll want to visit it any more. For Irish isles wildness and rough pastures are of the essence.

Here you get an acute awareness of the still unspoiled beauties of the west, an experience no other can match.

Eagle Island

No birdie avair soar any wing to Eagle
James Joyce

Eagle Island is named after the Golden and Sea Eagles which used to haunt it.

Otway's book *Sketches in Erris and Tyrawley* contains many stories of these kings of birds which were particularly numerous in Mayo.

There was the man who had fresh herrings every day by climbing down to an eyrie, putting a string round necks of the eaglets and removing the fish brought by the parents, well, there wasn't much else to eat on Eagle Island so we can forgive the theft and admire his bravery.

Another tells of a baby about to be carried off in its swaddling clothes by an eagle which then found it could neither rise with the load or disentangle its feet from the wool. The gallant mother killed the bird with a loy!

If they saw a sheep standing near the edge of a cliff two or more eagles would combine to drive the creature over the edge, and so found a feast at the bottom.

The coming of the French in 1798 and their handing out muskets to likely lads before departing, put guns into the hands of peasants who used them to wipe out the wild red deer as well as many larger birds.

Eagle Island lies at a key turning point of the coast so a lighthouse was erected in 1835 at a height of 220 feet above the sea. Even at this height the light keepers families were threatened by heavy spray and flying debris in strong gales. The west wall had to be heightened by twenty feet but white water still came over.

There are no signs of monastic buildings but it seems almost certain that the remarkable cluster of a dozen eremitic oratories and churches of the 6th century a few miles south would have had an outstation for brothers who had misbehaved, or wished to meditate. Their shelter would have been a turf or wooden bothy or a crevice in the cliff, leaving no long term trace.

The Irish were not the first to favour insular sites, St. Devanus had established a monastry on Ramsay Island in Wales in 182 AD, but they soon made up for being late starters. An unknown Irish monk wrote the reasons why islands were spiritually best.

'In the calm of summer the broad expanse of the ocean lay still and unruffled, mirroring in its blue depths the over-canopying heavens. Was it not a fair image of the unbroken tranquillity to which the heart of the recluse aspired? In the gloomy winter nights when great crested waves rolled in majestic fury against the granite headland could not the driving stormwrack remind the inmate that he had chosed the safer path by abandoning the world of passions which wreak destruction far more appalling?'

Jack Hawkins from Belmullet told us that in his time as a keeper, he was relieved from rocky Scotchport by a four-man currach – which was often impossible. Keepers did four weeks on followed by two weeks ashore. "You could say goodbye to Christmas if you went out there in November" he recalled.

The island provides no real shelter but in moderate weather an anchorage can usually be found. My son Milo and I dropped two picks off its north side one year and scrambled up the rock face to explore. All we saw was the sad and abandoned quarters where keepers families lived. The light is now unmanned but the willingness of men and women to endure such conditions must have relieved the worry of many a navigator uncertain of his position after an Atlantic crossing and saved countless sailors' lives.

Erris to Sligo

*'The story of the enchanted islands
of the west is fresh as morning dew'*

St. John Gogarty

Leaving Eagle Island we start our voyage along the north coast of magnificent Mayo. Huge rock stacks like King Kong's ninepins greet the islander close to cliffs fronting miles of uninhabited moorland.

Off Erris Head is Illandavuck sheering 174 feet, like a sentinel at the west of well named Broadhaven. Five miles east, on the other side of the bay, is Kid Island, a craggy mile long mass. It looks like a great aircraft carrier, flat on top

The Stags of Broadhaven

with sheer cliffs all round. Borowski, a famous 18th century rapparee, had a stronghold there with easily defensible one man paths the only access. The treasure he buried, like that of the island's buccaneer namesake, has not yet been found!

A mile on are the first of three pinnacles – The Parson with a silhouette like a robed preacher. A hundred feet high, it is dwarfed by its giant neighbours, The Hag (230 feet) and The Buddagh (260 feet), its name meaning 'the rogue'.

These three are dramatic to behold but too steep for grazing and deadly to sailing ships which could be pushed inshore by vast heaves of a swell if there was no wind to fill their sails. Rounding them some years ago after the big gale, which usually blows here at the end of August, we watched twenty foot waves rise to break on The Buddagh in white masses that flew up and crawled back. The colossal swell boomed with a deep hollow roar in the huge cave of Doonivalla a quarter of a mile beyond.

Abreast of these The Stags of Broadhaven, two miles offshore, loom black and bold. They comprise, within a half mile extent, four gigantic pyramids. Their heights in feet from south to north are – Teach Mor, 302; Teach Donal O'Clerigh, named for one of the Four Master Annalists, 306; Teach Beg, to the east, 233 and at the north An–t-Oighean 246.

The sounds between them are clean and there are decorative caves and tunnels. It is hard to think of another quartet so isolated and magnificent. The Witch's Hats which we sailed past another year off Cape Ortegal in north west Spain might just rate. I have seen sheep grazing on the Teach Mor but last time we passed in 2004 there were none. Shepherding is dangerous and time consuming, enough to deter even the hardy men of Erris. The difficulty of landing has preserved the Stags from occupation by monks, pirate or graziers.

I managed to scramble ashore once just north of the cave on the south side of the Teach Beg but saw no sign of a level space big enough to lie down on.

Even eagles could find no more than a single site – according to tradition a pair did nest year after year on the northern rock until shot by an enraged shepherd. The Stags will always be a magnificent sea-mark. The name is of course a corruption of 'rock stacks' but when first sighted they do look like the antler tips of a pair of Royals, bayed in deep water. And so I think of them!

The wildness and loveliness of these rocks is infinitely soul-satisfying, they look perfectly content to be uninhabited.

Pig Island or Inishmuc, counted good for grazing in old days, is named from a remarkable resemblance to a recumbent boar. It illustrates the grandeur of the cliffs behind by the way in which they dwarf its 200 foot height. The striking feature is the 60 foot high tunnel which Ros has drawn. A hundred years ago Coast Guard cutters could have been seen sailing through under scrubbed white sails. I have never had the right conditions of high water and low swell to try.

Illaunmaster, also known as Moista or Puffin Island, the next east, is a 300 foot high grassy cone. Its graceful shape and the verdant green come easy on the eye after miles of dark jagged rocks. Now a Nature Reserve, its earthen slopes

The Buddagh off Portacloy

Kittiwakes

provide ideal nesting sites for puffins who like a burrow in which to lay. Groups of these lovable birds were soon thick around us on the water, the first we'd seen in numbers in Connaught. They dive with a comical upturn of the tail. If occasionally this occurs close to the bow you can see them gliding swiftly away under water. Puffins are the 'Playboys of the Western World'. Their white chest, black neck band and red feet

Puffins on Puffin Island

give the appearance of a man who has changed into his dinner jacket but forgotten to remove his carpet slippers.

Their population has taken a beating lately due to the netting of sand eel for sale to mink breeders in The Baltic. At one time I saw forty boats setting out from Stornoway equipped for this disastrous hunt which removes the primary source of food for many species of sea birds. I think it is better controlled now.

A pair of peregrine falcons and several species of gulls also nest on Illaunmaster.

The Sound inside is dramatic, straight as a gun barrel and

two hundred yards long with high vertical sides. It looks like the Corinth Canal and is formed by what geologists call a trap dyke – that is a perpendicular vein of hard igneous material dividing the strata of the mountain. Inserted between, but not bonded to, the rocks on either side the insert has a tendency to subside or disappear over a period and leave a gap. A vivid example of the processes of evolution and decay.

It was not an occasion to take *Agivey* through but I hope one day to explore it properly. Illaunmaster is a place for bird lovers who are also good climbers to visit in the company of a Warden.

A few miles east Downpatrick Head extends like a wedge with its base to seaward, its face a serried row of caves.

A drum-shaped 200 foot stack, as high as the cliffs, stands about a hundred yards to seaward. It is named Doonbristy (the broken-off fort). In the 5th century it was part of the main promontory and held a church of St. Patrick. Erosion,

Black Guillemots

plus an exceptional gale, caused a rock fall that left what is now the stack in isolation and the church building split. You can see remains of its wall from sea level. Local rumour had it that the part on the stack was stuffed with gold! This was the loot of a 'paganee king' who was using it as his stronghold. The monks are said to have tried to get up the stack by cutting steps. Traces of a few artificial toeholds on the lower part are still to be seen but nothing like a staircase. It took a Danish skipper to get there – with a kite. He managed to float a light line over the top, pull up a stronger one of heather and make the ascent.

If he found anything he kept it to himself.

Nowadays a climber able to lead on 'Very Severe' would get with up with the use of a few pitons. Doonbristy is banded in layers of dark grey and blue limestone, yellow sandstone and a rusty iron-bearing conglomerate. All are tilted with the upper end to the north, matching those of the adjacent cliffs. White breasts of kittiwakes and guillemots line the ledges.

Rowing round its soaring pillar on a calm day, beside the overhanging mainland cliffs and nosing into a cave or two is strangely impressive, something unusual in most island visitors' experience. The sense of danger is increased by the menacing way that waves coming in from the west curl right round the east side of Doonbristy. We could see them boomerang back and break across the Sound which is known as the Giant's Leap.

There is always a swell on the west coast of Ireland. Except when there isn't. That's only a week or two each summer.

So you must pick your time. You may be sure big rollers are roaring round Doonbristy as you read these words.

The coast changed its nature as we crossed the low open bay of Killalla (emphasis on the second syllable). We could have found shelter there behind sandy Bartragh Island where a new championship course was under construction, but lacked a set of golf clubs. So we sailed on past the low limestone shore of Roscommon, a strange contrast to Mayo cliffs.

In a couple of hours mountains of a different type rose slowly higher. Saw-edged Benbulben keeping guard over the Yeats grave to the north and the round hump of Knockrea to the south marked the entrance to Sligo.

In there would be the relief of shelter. Uninhabited isles, however dramatic to behold, are never as attractive as those on which people live like Coney Island now close ahead.

Pig Island Arch

Doonbristy

Yeats Country Islands

'Sailing, sailing to the County of Mayo'

Yeats

Sligo Harbour is handsomely situated with the sheer inland cliffs of Benbulben to the north and, as an old Admiralty Pilot puts it,

'the very remarkable solitary limestone mountain of Knocknarea (Hill of the King) *to the south.'*

Its summit is crowned by an ancient tumulus a thousand feet above sea level.

That pimple contains forty thousand tons of boulders and commemorates Maeve, Queen of Connaught.

The harbour is five miles long by two wide. All the south side dries out, so at low water the navigable area is confined to a narrow strip along the north shore. This is fortunate for, if deeper, the bay would often be packed with ships on moorings and disfiguring fish farms. Unsullied today, its silvery waters reflect the mountains to make it the most beautiful bay of its size in Ireland.

On a map the bay is shaped like a lion's maw. Coney Island might be said to resemble a lump of tasty meat into which the lion is about to get its teeth.

Coney is flat and sandy, a mile long by three quarters wide, rising no more than 50 feet in a few dunes at the south. It

is the source of much history and myth. In the late 19th century the waters round it were something of a playground for Jack Yeats, the painter, who spent summer holidays in Elsinore House at Rosses Point. His Pollexfen grandfather ran a shipping line so could arrange expeditions in company boats. Brigs, barques and schooners sailed or were laboriously towed up the river past Coney at a time when 500 ships a year brought cargoes to Sligo. Early steamships chugged in and out in the Channel where 'the man who never told a lie' points the way past the shallows. This is the famous effigy of a naval petty officer, known as the Metal Man. He stands pointing to the safe channel on a rock where the shipmasters could become confused at the fork either side of Oyster Island.

On a boat trip with Doctor Des Moran on a chilly March day 2005 we first passed the harmonic hump of Oyster Island, sausage shaped and half a mile long. It has a graceful white lighthouse tower at the seaward end and in the middle a fine square of stone walls, which were painted white in days past by Irish Lights as a day mark. There are traces of an early church. Horses had the pale March grass eaten down flat but a flock of herons were making a good meal on the foreshore.

We landed beside Coney's large stone pier, well protected by rock armour to the west. The old fort on the north east point provided elevation to watch the dangerous currents of Shrunamoyle in the quarter mile narrows between where we stood and Oyster Island. As the ten square mile bay drains through, tidal streams swirl faster than you could walk with eddies and tide rips which only local men can understand. The Admiralty, in a rare comment, labels it 'Dangerous Passage'. Experienced kayakers love to play in the white water.

At times it was a place to muster armies about to raid the interior. Coney is associated with St. Patrick and has a rocky seat named after him.

Across rolling grass south of the pier is a row of a dozen cottages. A public house with good Guinness and curious curios is at the centre. The Magowan family, kings of the island and harbour pilots, have run it for generations.

We were made most welcome in the only occupied house by George Sleater and his partner Gabrielle. Thawing out by their log fire we admired his books, pictures and relics including a Humane Society Award to his grandfather for saving three men from a stranded ship in huge seas.

Then along a muddy track we explored the west side where the swell roars endlessly in winter on Deadman's Point. The Cluckhorn, a mile long stone barrier made a fine sight rimmed with surf. It connects Coney to Black Rock which carries a 60 foot white lighthouse. The boulders were planned to divert the tide so that it would scour the channel and save dredging. It didn't do that but its endurance is a tribute to the famous port engineer Nimmo and still provides shelter for a boat to anchor beside Black Rock, a pleasant place for a picnic at half tide.

You can get to Coney overland from the south near low water, following a row of stone beacons but be careful.

The Metal Man, Sligo

People have been drowned and vehicles trapped by the rising tide which covers the road four feet deep at springs.

An islet half way across is named Doonanpatrick after the Saint. Though barely thirty yards across at high water it has signs of a kitchen midden, evidence of occupation by fish-eating Bronze Age folk a thousand years before the days of St. Patrick.

One comes away with an impression of Coney as an historic island where the mystery of the sea and the fire of a sunset weave a magic spell.

As we jumped ashore on rough muddy gravel below Wheatrock, his lovely modern house, Des remarked, "Isn't it great not to grow up?"

That exactly echoed my feeling. We had enjoyed that short island trip as much as we would have done as teenagers.

Postscript

'The isles are lovely, dark and deep,
But I have promises to keep'.

After visiting a cocktail of islands it is amusing to try to define the ideal one.

Walter de la Mare, the poet, listed;

'a creek, a spit of quicksand, dunes, caves, precipices, cataracts, streams, an unfathomable lake, and to be all but inaccessible.'

R. M. Lockley, who lived on Welsh Skokholm, adds as essentials;

'A copse, birds and some agreeable human company'

Each omits a harbour or at least slipway which I would want for a boat.

Irish isles have all these items but not in one place. They have brought us great benefits. Estyn Evans, the distinguished geographer, reminds us that Ireland's first taste of civilisation and first knowledge of metals came from prehistoric seafarers making their way from island to island. So a coast needs inhabited isles. A decline in population started in the 1930s. This came mainly because of the need for secondary education. It got quicker about 1960.

Recovery started in the late 1990s when the RIB (Rigid Inflatable Boat, with outboard engine) made commuting easier and demand for coastal housing increased. County Councils began to pipe out water and electricity. One hopes to see this continue so that island communities will prosper and increase.

Ros and I wish you well with your island visits.

For myself, as my Donegal friend Griff O'Donoghue wrote,

'No isle, I know, will hold me long
When the sea sings again its syren song'.

Dear Reader

This book is from our much complimented illustrated book series which includes:-

Belfast	Blanchardstown, Castleknock and the Park
By the Lough's North Shore	Dundrum, Stillorgan & Rathfarnham
East Belfast	Blackrock, Dún Laoghaire, Dalkey
South Belfast	Bray and North Wicklow
Antrim, Town & Country	Dublin's North Coast
North Antrim	Limerick's Glory
Inishowen	Galway on the Bay
Donegal Highlands	Connemara
Donegal, South of the Gap	The Book of Clare
Donegal Islands	Kildare
Sligo	Carlow
Mayo	Armagh
Fermanagh	Ring of Gullion
Omagh	Carlingford Lough
Cookstown	The Mournes
Dundalk & North Louth	Heart of Down
Drogheda & the Boyne Valley	Strangford's Shore

For the more athletically minded our illustrated walking book series includes:-

Bernard Davey's Mourne   Bernard Davey's Mourne Part 2   Tony McAuley's Glens

Also available in our 'Illustrated History & Companion' Range are:-

Ballymoney   Banbridge

And from our Music series:-

Colum Sands, Between the Earth and the Sky

We can also supply prints, individually signed by the artist, of the paintings featured in the above titles as well as many other areas of Ireland.

For details on these superb publications and to view samples of the paintings they contain, you can visit our web site at **www.cottage-publications.com** or alternatively you can contact us as follows:-

Telephone: +44 (028) 9188 8033   Fax: +44 (028) 9188 8063

Cottage
Publications

Cottage Publications
is an imprint of
Laurel Cottage Ltd
15 Ballyhay Road
Donaghadee, Co. Down
N. Ireland, BT21 0NG